SCIENCE IN GOVERNMENT

J RONAYNE

EDWARD ARNOLD

© Jarlath Ronayne 1984
First published 1984 by
Edward Arnold (Australia) Pty Ltd
80 Waverley Road
Caulfield East Victoria 3145

Edward Arnold (Publishers) Ltd
41 Bedford Square
London WC1B 3DQ

300 North Charles Street
Baltimore MD 21201 USA

National Library of Australia
Cataloguing-in-publication data

Ronayne, Jarlath, 1938-
   Science in government.
   Bibliography.
   Includes index.
   ISBN 0 7131 8040 4.
   1. Science and state.   I. Title.
500

Designed by Peter Yates
Cover design by Lauren Statham
Typeset in Baskerville by Setrite Typesetters, Hong Kong, and printed in Singapore by
Richard Clay (S. E. Asia) Pte Ltd

Contents

To Ros, Alex and Tim

Preface

This book originated in a request by the Australian Science and Technology Council (ASTEC) to review the literature on assessing priorities in research and development and to analyse the criteria, techniques and mechanisms used to enact science policy in selected OECD countries. ASTEC advises the Prime Minister on priorities and research activities generally, and it was intended that the results of the study would be used to develop criteria and techniques that would assist ASTEC in this task. The two elements of the study provided the framework for a book on the principles and practice of science policy, and after my report had been accepted and circulated for comment in Australia and overseas, the Council kindly gave me its permission to use it for this purpose. I used a subsequent period of secondment to the ASTEC Secretariat to extend the analysis and to bring the reports on individual countries up to date for publication.

It has been suggested that in science policy we have been taking one step forward and two steps back, and it has to be acknowledged that the main problem addressed in my report to ASTEC — that of establishing workable criteria for deciding which areas of research and development will receive government funding — has remained intractable. But I have tried to show in this book that there has nonetheless been real progress in the theory and practice of science policy.

In Chapter 1, science policy as a notion is traced from its origins in the seventeenth century through to the twentieth century, when the notion became a reality and governments began to intervene in the research system. In Chapter 2, the arguments for and against government intervention are discussed in the light of science policy studies that have transformed our view of the relationship between science, technology, innovation and the market. The conclusion is reached that government intervention is justified, and in Chapter 3 the criteria for identifying priority areas in research and development and the mechanisms for enacting science policy are considered.

Most of the empirical work on criteria for choice has centred on the support of pure research. In the case of strategic and applied research, especially that carried out in government agencies, it has generally been assumed that a nation's priorities are decided by the political process and that the research system is then set in motion to tackle the problems posed in the most effective way. In this area, therefore, interest has focused on the science policy machinery that might best be established to assist governments and their scientists and technologists to work together fruitfully and harmoniously. Four types of science policy systems are identified, and in Chapters 4 to 8, the systems in operation in a number of OECD countries are reviewed and loosely classified as one of these types. The reports on individual countries are not intended to be exhaustive, since excellent detailed studies of most of these countries are already available. Australia was not included in the original ASTEC brief, but I have included it in this book and hope that the discussion will be of interest to international as well as Australian readers.

Finally, in Chapter 9, the ways in which governments are trying to reorganize their research systems to meet the challenges posed by what is referred to as the new economic and social context are reviewed.

This book is intended as a text for students and staff on courses with a science policy studies component. It should also be of interest to national science policy advisers and their secretariats, public servants involved in formulating and implementing science policy in ministries and departments, policy advisers in scientific institutions and, finally, practising scientists, who are likely to be subject more and more to the co-ordination inherent in science policy systems.

I would like to express my sincere thanks to Dr Roy Green, Deputy Secretary of the Australian Department of Science and Technology, who, when Secretary to ASTEC, gave me the opportunity of spending a rewarding and stimulating period as a member of the ASTEC Secretariat. Special thanks are due to Dr Bruce Middleton, now Secretary to ASTEC, who directed the priorities project and whose help and collaboration I valued greatly. Finally, I would like to thank Mrs Ailsa Bailey, who typed the entire manuscript and provided invaluable assistance to me at every stage of the project.

J. Ronayne
Sydney 1983

CHAPTER 1

Knowledge and power

A brief history of science policy

The governments of most industrialized and industrializing countries now acknowledge that they must support scientific research and development (R & D) not only in areas that are potentially relevant to the work of their administrative departments and agencies but also in areas of research that seem to have no prospect of either immediate or long-range application. Governments give massive financial support to scientific research in universities, other institutions of higher learning and private research foundations. In the private sector, businesses receive direct subsidies, tax advantages and other financial incentives to encourage them to invest in R & D. And in government establishments, R & D is carried out which is directed towards the needs of government departments and agencies and towards the social and economic benefit of the nation as a whole.

The United States provides a striking example of government commitment to scientific research. In the 1983 financial year the Federal Government's appropriations for R & D amounted to US$43 billion (1 billion = 1000 million). The Department of Defense, for example, was allocated an R & D budget of US$24 billion and the National Science Foundation, an agency with no laboratories of its own, being generally concerned with the support of R & D in external institutions, received more than US$1 billion. In 1939, the United States Government spent less than US$50 million on scientific research and, at the beginning of this century, it spent less than US$5 million. The governments of many other industrialized nations have increased their research expenditures at a similar rate. In the United Kingdom, for example, the central government's contribution to what is now called the Gross National Expenditure on Research and Development (GNERD) in the 1981 financial year was more than £2 billion. In 1937, it spent about £3 million.

Why did scientific research become recognized in the twentieth century as a legitimate area of government responsibility, and why did government expenditures on research rise at such a phenomenal rate, especially in the second and third quarters of the century? The answer is that in the twentieth century, and especially in the second and third quarters, the connection between scientific knowledge and economic and political power was finally demonstrated. As a result, the concept of science policy was born. Science, like economics and foreign relations, became an area of government concern, and there are few countries in the world that do not now claim to have a national science policy and the machinery to enact it.

A rational and comprehensive science policy has been defined by Shils as 'the intention to influence scientific development through authoritative decisions.'[1] Broadening the scope of this definition, a rational and comprehensive science policy will be defined here as the intention to influence the development of science and technology in a coherent way by authoritative and informed decisions in order to fulfil national economic, social and political objectives.

A comprehensive history of the notion of science policy is beyond the scope of this book, but some of the key concepts and concerns can be illustrated by referring to some early thinkers' ideas about the relationships that could and should exist between science and government.

It is customary to trace the origins of the notion of science policy to Sir Francis Bacon, the seventeenth-century lawyer, essayist and Lord Chancellor of England during the reigns of Queen Elizabeth I and King James I. Bacon recognized the promise that 'natural philosophy' held out for what he called the relief of man's estate. 'Knowledge and Power meet in one,' he said, 'for where the Cause is not known, the Effect cannot be demonstrated'. In his utopian essay *New Atlantis*, published in 1627, he set out his vision of a society in which natural philosophy, organized and planned, would be the people's servant and an agent of social change. In this utopia the central and most important state institution was Salomon's House, an imaginative forerunner to our present-day R & D establishment, wherein all kinds of scientific research, development and demonstration were carried out by the Father of the House and the Fellows. Results that could be of practical benefit were disseminated to the wider community. Those Fellows who conducted new experiments—'experiments that they themselves think good'—were called Pioneers or Miners; and those who applied the results—'looking into the experiments of their fellows and cast about how to draw out of them things of use and practise for man's life'—Bacon called the Dowry-men or Benefactors. The Depredators were 'those who collected experiments that are in all books', and the Inoculators executed experiments that were decreed by higher authorities, the Lamps. All of the Fellows were supported by novices, apprentices, servants and attendants.

The state supported Salomon's House because of the usefulness of its work. Bacon, in all his writings, emphasized the practical utility of natural philosophy. In *New Atlantis* he says, 'The end of our Foundation is the Knowledge of Causes and secret motions of things; and the enlarging of the bounds of Human Empire, to the effecting of all things possible'. Bacon recognized that 'natural philosophy' could be an instrument of government policy, and he created an albeit imaginary organizational structure within which science and the state would flourish.

Viewed from a modern perspective, Bacon's organizational structure raises more questions than it answers. Bacon assumed open-ended support for all the scientific activities that took place within the walls of Salomon's House. Resources were never scarce in his imaginary research establishment and the Fellows, it appears, did not have to compete for their share of the financial cake. No one seemed to be responsible for allocating resources, and priorities were never set. And apart from a primitive vetting system in which Fellows decided which discoveries they would release to the state and which they would keep secret, the body corporate appeared to be quite oblivious of the problems that were likely to arise when a creative activity that held such potential for acquiring power over nature and humanity was supported by external agencies.

According to Lakoff, however, Bacon created a climate of opinion in England that was, at least, favourable towards science.[2] He helped gain acceptance for science from those who might otherwise have reacted against it. He recognized the power of the 'New Learning' and he recognized, too, the need for state support of science, both moral and financial. However, knowledge and power were to remain stubbornly apart for more than 250 years after Bacon's death in 1627. The Royal Society of London, founded in 1646 as a result of the inspiration provided by Bacon's writings, received financial support from King Charles II, but this support was both meagre and short-lived—despite the polemics of its historian Thomas Sprat who, in the words of Stephen Toulmin, could 'write as persuasively as Jerome Wiesner [President Kennedy's science adviser]—and in even more sonorous language—about the technological "spin-off" to be expected from this royal patronage.'[3]

In the eighteenth century the French aristocrat Antoine de Condorcet was a brilliant early exponent of the notion of science policy. Perhaps because of the times he lived in, Condorcet was more aware of the problems Bacon failed to address. He was an active revolutionary, and the scientific society he envisaged was democratic. Whereas Bacon's New Atlantis was an island monarchy, Condorcet's scientific utopia was a republic, with science supported by an enlightened populace governing itself according to the principles of reason. He argued that, if history was any guide, a monarch could be expected to support only those projects that took his fancy or flattered his own vanity; and under a monarchy there could be no guarantee of continuity of support. Within a democracy continuity of support would be more likely, and there would be less abuse of power and privilege in science affairs as in other affairs of state. Though the natural philosophers would be accorded a special place in the democracy, they would not by virtue of their knowledge be the rulers, nor would pomp and circumstance attend them in the manner of the Father of Salomon's House, to whom Bacon accorded almost regal status.

Condorcet's statements on autonomy, accountability and the allocation of resources, issues which concern us to this day, are quite remarkable when viewed from the perspective of the twentieth century. The natural philosophers, through an elected council, would be responsible both to the national electorate and, at a higher level, to the civilized nations of the world. The council would have few but far-reaching powers and be autonomous within its own domain. One limitation placed upon the council is evident from Condorcet's proposals concerning resource allocations:

> It is for the association of savants [natural philosophers] to judge independently what it believes should be undertaken to speed up the progress of the sciences. It is for the public authorities to judge with the same independence which of its projects merit support or even munificence. Let us not despair of seeing the time when this division can be effected through reason alone.[4]

Condorcet was, then, very conscious of the need to preserve the independence of the scientific community from the state but he left the decision on which projects should be funded to the public authorities. Though he stipulated that these public authorities should be enlightened enough to feel that 'they must not direct the work but second it', he did, in the end, ask them to direct it. He insisted that state support for science could be successful only if the scientific institutions set up as a result enjoyed sufficient autonomy to regulate their own affairs, yet he interpreted this autonomy as the freedom to accept or reject state support, while the state would be free to supply funds only as it saw fit.

Salomon, in explaining Condorcet's attitude towards autonomy and state support of science, suggests that he was not contemplating the possibility that scientific work would be so costly as to demand massive and almost complete state support.[5] Condorcet's scientific community would live mainly on the members' contributions, with a reserve fund that would provide income for its larger needs. In this scientific utopia, the state would be neither the scientists' primary source of support nor the prime customer for their products. The state came into Condorcet's scheme only to guarantee financially the continuous development of research. Whereas Bacon recognized that the pursuit of science for itself was a 'more excellent and fairer thing than all the uses of it', it was as an instrument of government policy that he argued for its public financial support. Bacon's ideas reflected what we now call 'science in policy'. In Condorcet's scheme of things, science contributed to social progress not so much by its application but more by the purpose it was to serve, and its purpose was enlightenment. It served the state by emancipating the minds of all free men. Condorcet's policy was mainly a policy for the advancement of science or, in the current science policy usage, it was a 'policy for science'.

Though Condorcet advocated a policy for science, his fellow revolutionaries, and eventual executioners, saw science in a different light. The revolutionary Committee of Public Safety quickly established a committee of scientists, the *Congrès des Savants*, which advised the government on matters of state to which science could make a contribution. A Committee of Four Citizens learned in chemistry and mechanics was established to seek out and prove new means of defence. So, for the first time in history, science became an instrument of government policy and scientists acted as advisers to a government. However, as Salomon points out, though science and scientists entered the corridors of power during the French Revolution, a rather sinister precedent was set, because they also became involved in a situation where their allegiance to a regime became the dominant factor in deciding whether or not they would receive government patronage and support. Scientists who worked for the Revolution were chosen as much for their political affiliation as they were for their special knowledge. Past associations with previous regimes could be fatal, as Lavoisier, the great aristocratic chemist, was to find.[6]

The establishment of a scientific advisory council and elevation of scientists to a position of power in the First Republic could have been the beginnings of sustained governmental support for science and the institution of a national science policy. But it was a flash in the pan, despite the work of Condorcet and the other *philosophes* of the time. As a result of a suggestion by Condorcet, the National Institute for Science and Art was reorganized in 1794 and the famous Ecole Polytechnique was founded in Paris in 1796 for the scientific education of military officers and engineers. For the first time, therefore, science was taught at a high level in its own right in well equipped institutes of higher learning. But consistent support for science, and the machinery to effect it, remained a mirage for some time to come.

The emphasis in post-revolutionary France was on short-term military needs. The *Congrès des Savants* enjoyed prestige because its work contributed to military technology and hardware, not because it contributed to the advancement of science itself. But at least its work was beginning to demonstrate the link between knowledge and power. The spin-off from government patronage, together with the educational reforms of the early nineteenth century, resulted in France's scientific ascendancy in Europe until close to the middle of that century. The Collège de France, founded as the Collège Royale in 1626 by Louis IV, the Ecole Polytechnique and the Museum d'Histoire Naturelle became leading scientific institutions and by the end of the First Empire set the pattern of French scientific, technical and educational organization which was to last until the middle of the twentieth century.

Science and policy in the nineteenth century

The immediate result of the savants' success in reaching the corridors of power in France was to encourage the cultivators of science, as the natural philosophers were by then called, to press for government support for science in their own countries. In eighteenth-century England one of the most vociferous exponents of government support for, and encouragement of, science was the mathematician and inventor Charles Babbage. Babbage, Lucasian Professor of Mathematics at the University of Cambridge, in his famous essay 'On the Decline of Science in England', which he wrote in 1830, castigated the British Government for its lack of support in terms that are now familiar to us. He said:

> It cannot have escaped the attention of those whose acquirements enable them to judge, and who have had opportunities of examining the state of science in other countries, that here in England we are much below other nations, not merely of equal rank, but below several even of inferior rank. That a country 'eminently distinguished for its manufacturing ingenuity' should be indifferent to the progress of inquiries which form the highest department of that knowledge on whose elementary truths its wealth and rank depend is a fact which is well deserving the attention of those who shall inquire into the causes that influence the progress of nations.[7]

Babbage makes two important points in this statement. First, a case is made for increasing the level of scientific research in England on the basis of international comparisons and, second, it is assumed that manufacturing ingenuity depends on 'elementary truths' derived from scientific knowledge. According to Babbage, the remedy for Britain's poor showing in the international league tables was government intervention in the research system. He recommended that the status of the scientific researcher should be enhanced by awards and honours, which at the time were the preserve of the aristocracy and the captains of industry who, incidentally, were doing quite well in England without the very truths that Babbage considered to be essential to their productive activities, and also that the government should support scientific research financially. Babbage was farsighted enough to realize that if the prime justification for carrying out scientific research was economic — that is, if the technological benefits arising from it were the main concern — then there were good reasons for governments not to support it. First, he said, the public who consume the technological products arising out of scientific research, or profit by the invention and diffusion into the market place of such products, are far better judges of their merits than any government; in other words, market forces should prevail in the decisions as to which research should be carried out, and those who are embedded in the market system and rely on it for their livelihood should initiate and financially support the appropriate research. Second, if the research is successful and the invention subsequently appears in the market place, then the benefit that arises from the sale of the product is much larger than any amount that a parsimonious government might bestow, and this should serve as sufficient incentive for the entrepreneur to invest in appropriate scientific research.

Under this system, however, Babbage considered that all 'abstract truth', that is, pure science, would be excluded from reward. If it is important to the nation that abstract principles should be applied to practical problems, then it is clear that, in Babbage's own words, 'encouragement should be held out to the few who are capable of adding to the number of those thrusts on which such applications are founded'. In justifying government support for pure science, he cited the long intervals that often elapse between the discovery of a new scientific principle and its application in a product or process, an argument that again presupposes that technical inventions and technological innovations generally depend upon prior discoveries in pure science. Pure science was, therefore, as economically important, if not more so, than the less abstract sciences and an agency should be set up to support it. Thus:

> Unless there exist peculiar institutions for the support of such [abstract] inquirers, or unless the government interfere, the contriver of a Thaumoscope may derive profit from his ingenuity, whilst he who unravels the laws of light and vision on which multitudes of phenomena depend, shall descend unrewarded to the tomb.[8]

Babbage's policy was a policy for science in the sense that we now apply this term. Pure science should be supported by the government because it is economically important, because in the absence of government support the pure researchers on whose labours the profits of the entrepreneur and industrialists rest go unrewarded, and because the nation has a responsibility to maintain a pure research effort commensurate with its economic and political status. Unlike Condorcet's policy for science, which was moderated by the demand that scientists should be accountable to a higher authority which should play a dominant role in deciding which projects should be funded by the government, Babbage appeared to have only one priority, the advancement of science and of scientists. He was prepared to use an as yet undemonstrated connection between the advancement of science and economic benefit to achieve it.

The case for government intervention in British science was later taken up by Alexander Strange, an admirer and contemporary of Babbage. In 1869, Strange proposed to the British Association for the Advancement of Science, a body set up by Babbage in 1836, that it 'undertake an enquiry into the question of whether there existed in the British Isles sufficient provision for the prosecution of Physical Research and, if not, what measures were needed to secure it'. The Association appointed a small committee to look into the matter and, concluding that the provision was far from adequate, it recommended the establishment of a Royal Commission to consider:[9]

- the character and nature of existing institutions and facilities for scientific investigation, and the amount of time and money devoted to such purposes; and
- what modifications or augmentations of the means and facilities that are at present available for the maintenance and extension of science are requisite and in what manner these can be supplied.

Evidently the government held the British Association in sufficiently high regard to take notice of its recommendations because a Royal Commission was duly established under the chairmanship of the Duke of Devonshire. In a series of reports written between 1870 and 1875, the Devonshire Committee came to the view that scientific education needed to be encouraged and that state institutions needed to be established to carry out scientific research and to allocate funds to researchers.

In its Eighth Report, published in 1875, the Committee considered the mechanism by which the government might achieve this. Its recommendations included:

- the establishment of a special council representing the 'Scientific Knowledge of the Nation', nominees of the government, the Royal Society and other scientific societies;
- the creation of a Ministry of Science and Education.

This was the first time that a special Ministry for Science was advocated. The reasons put forward will be familiar to those who have been involved in or who have studied the events surrounding the establishment of such ministries in the twentieth century. In his evidence before the Commission, Alexander Strange had this to say:

It seems to me that in the first place there should be some means of bringing science fully before the nation through Parliament. I know of no means of doing this that is in accordance with our constitutional procedure except through a Minister of State; and therefore assuming science to be a matter of enormous national importance, I think it is essential that it should all be brought under one Minister of State who should be responsible to Parliament for everything which is done in the name of the nation to advance science, and who should frame his own estimates and keep them distinct from those departments which have little or nothing to do with science.[10]

In the writings of Bacon, Condorcet and Babbage the notions of a policy for science and science in the furtherance of policy have now been identified. With Strange, and the Devonshire Commission's Report, a new element was introduced — the notion of special science policy machinery within government. As we shall see in Chapter 3, Strange proposed a centralist mechanism with the nation's scientific activities concentrated in one Department of State, under the political control of a Minister, who would frame his own 'Science Budget'. A former professional soldier, Strange exempted defence science from this arrangement but suggested that there should be a mechanism for bringing the military's needs to the attention of the civilian scientists. The proposed Council of Science would, he suggested, perform this task.

But, as MacLeod has noted, Britain in the 1870s and 1880s was preoccupied with matters for which science could offer no immediate solutions — imperial expansion, economic instability, an inadequate education system, poverty and unemployment.[11] It was unlikely, therefore, that scientific education, research and policy would receive high priority, and for twenty-five years after the publication of the Devonshire Commission's final report there was no decisive action. Then, in 1900, the government established a National Physical Laboratory (NPL) to standardize and verify instruments, test materials, determine physical constants and undertake scientific investigations which were of great importance to the country's *industrial* interests. In a country that had been content until then to leave the nation's industrial research effort to the mercy of free market forces, this was a remarkable innovation, though it had been advocated constantly from Babbage's time onwards.

To understand the reasons for the establishment of the NPL, we need go no further than the Report of a Treasury Committee set up in 1898 to consider the desirability of establishing such a laboratory. The Committee issued a favourable report and recommended that a National Physical Laboratory 'should be established at the national expense on lines similar to, though not yet at present on the scale of, the Physikalisch-Technische Reichsanstalt.'[12] Leadership in industrial research had long since passed to Germany, and it was clear from the evidence submitted to the Treasury Committee that this leadership was due to government support and better organization of German industrial research. Moseley, who has chronicled the origins and early years of the NPL, says that the stimulus for change in the government's attitude towards industrial research support came from three sources. First, there was a justifiable fear of foreign, especially German, industrial competition; second, Britain's share of export earnings in the world economy was declining; and third, there was concern over the administrative and military failures of the Boer War.[13] The German Government's superior system of scientific and technical education and its support for industrial research were not lost upon the leaders of the British scientific community, who, in turn, ensured that the British Government was made aware of them.

The National Physical Laboratory represents the first real step taken by the British Government to intervene in what we will now call the national research 'system'. Until 1914, it was rather a small step. The total government grant to the National Physical Laboratory between 1900 and 1914 was about £100 000. In 1914 the First World War broke out, an event that boosted scientific research in Britain to an extent that an economic crisis could never have done. Salomon has said that the originating motives for 'policy through science' have always been military; in Britain this was not the case, but it took a war to make sufficient funds available to implement such a policy.

Science, technology and war

At the beginning of the twentieth century, science and technology played only a minor role in military affairs. Military hardware had, of course, been improved progressively during the nineteenth century. The use of railways for mass troop movements, the field telegraph in tactical deployment and the introduction of the machine gun are some examples of technological changes that altered the techniques of war during that time. But these 'improvements' had little to do with the systematic application of scientific knowledge. More often than not it was artisans, independent inventors, craftsmen and military engineers who effected the changes. Though 'scientists' were employed in military arsenals, their work tended to illuminate the practice of science rather than the 'art' of war. Count Rumford's brilliant work on the mechanical equivalent of heat, stimulated by his observation of the great amount of heat generated in boring cannon, was carried out in a Prussian arsenal. Lavoisier's work as head of the Régie des Poudres, the French arsenal, contributed more to the revolution in chemistry than it did to the French Revolution. This situation changed in the twentieth century, and the link between knowledge and power was conclusively demonstrated to governments and the military during the First World War when the seeds of scientific war were sown. From then on scientists would be indispensable.

The First World War is sometimes called 'The Chemists' War', but this is a misnomer if it is meant to imply that the importance of chemistry and chemists was widely recognized by the belligerent powers or that the Allied victory was due to the decisive use of chemical weapons. It is true that, in Germany, the chemical industry was able to supply artificial nitrates to the munitions factories when supply of Chilean saltpetre, the cheap naturally occurring raw material for the manufacture of gunpowder, was cut off by the British Navy. And on the Allied side, the importance of science-based industry, and the extent of their technological dependence on Germany, was brought home when supplies of dyestuffs, medicines and optical glass, all highly strategic materials, were suddenly cut off. As Professor E. Frankland put it:

> Can anything more humiliating be imagined than that the greatest Empire which the World has ever seen should have found that some of the commonest and most important drugs could no longer be dispensed, that the uniforms of our gallant soldiers could no longer be dyed to a constant colour, that our laboratories were crippled and paralysed by the want of some of the most important reagents and pieces of apparatus?[14]

Despite the warning, however, the importance of science and scientists in the war effort was not quickly recognized by the governments or the military. Young scientists were sent to the front where many, including the brilliant Moseley (see Chapter 2), were to die. In the various armies, reactionary and outmoded thinking throughout the senior levels prevented the full use of scientific expertise until the war was well under way. For example, when a leading English physicist offered to organize a meteorological service for the army at the beginning of the war, he was told that the British soldier fought in all weathers. It was only after the great losses in the Flanders mud that such a service was launched. It is widely reported that, when America entered the war in 1917, the American Chemical Society offered its services to the Secretary of War and was informed by the Secretary that its assistance was not required as the War Department already had a chemist. Even at the end of the war, the Chief of Staff at the War Department felt able to claim that 'now as always heretofore, the infantry, with rifle and bayonet must bear the brunt of the assault and carry it to victory.'[15]

As the war progressed, however, increasing numbers of scientists were employed in military research, perfecting existing weapons and devising new ones. The new chemical weapon that is most commonly associated with the First World War is poison gas but, though this may have reduced the death toll in the last year of the war because it was not always as fatal as bullets, shrapnel and so on, it had little effect on the outcome. It was both difficult and dangerous to use, and relied for its success on fairly accurate weather forecasts. It eventually became clear to the strategists on both sides that decisive advantages could not be gained by the use of poison gas. In view of the relatively low funding and poor organization of scientific research during the war, all that could be achieved was that the technical innovations of one side were matched rather than outstripped by the other. Sometimes it was found that tactical modifications, like anti-gas training, could lessen the impact of the innovation in any case. So, despite the introduction of novelties such as poison gas, the tank, the aeroplane and the sub-marine, none of which conferred a decisive advantage on either side, the First World War was fought on classic lines. It was, as Sapolsky said, a clash of massed armies because most of the generals had no idea of how best to deploy such novel weapons.[16] H.G. Wells considered that, on the British side at least, the military had hardly even come to terms with the machine gun, invented some fifty years before. He said:

> Since this War began we have been piling up infantry recruits by the million and making strenuous efforts to equip them with rifles. In the meantime the actual experiences of the War have been fully verifying the speculations of imaginative theorists. The idea that for defence purposes one well-protected skilled man with a small machine gun is better than a row of riflemen is a very obvious one indeed, but we have disregarded it.[17]

Organizational initiatives in the United Kingdom

It has already been pointed out that Britain's dependence on German technology became painfully obvious when the First World War broke out. The industrial giant of the eighteenth and nineteenth centuries, the home of the Industrial Revolution, had lost its industrial leadership to a country which, because it had smaller colonial markets than either Britain or France, simply had to be innovative or perish. In the late nineteenth century it was recognized in Germany that future industrial strength lay with technological innovation and a unique educational system was established which emphasized practical training in experi-mental science, even for administrators, and research teams were organized around great men. By 1914, German supremacy in science-based industry was unquestioned. In Britain, this supremacy had been recognized long before the outbreak of war, leading to the establish-ment of the National Physical Laboratory. But the parlous situation that still existed at the beginning of the war convinced the leaders of the scientific community and some members of the government that further action was needed. After considerable lobbying by the scientific community, the government decided to establish a Privy Council Committee for Scientific and Industrial Research under the chairmanship of the Lord President of the Council (a Minister without Portfolio), which would be responsible for a new fund that would, in due course, be provided for scientific and industrial research. A small Advisory Council, composed mainly of eminent scientific men, and men engaged in industries dependent upon scientific research, was also set up to advise the Committee on

- proposals for instituting scientific research,
- proposals for establishing or developing special institutions or departments of existing institutions for the scientific study of problems affecting particular industries and trades,
- the establishment and award of research studentships and fellowships.

The President of the Board of Education (later the Ministry of Education) would answer in the House of Commons for the Privy Council Committee.

The Privy Council Committee and the Advisory Council were the first stage in the development of a British science policy machinery. Scientific and industrial research were to be centrally supported by a government committee, with an advisory council to the committee ensuring that scientists rather than bureaucrats determined the directions in which the funds would flow. The emphasis was on the promotion and organization of scientific research with a view especially to its application to trade and industry, but in its first Report the Advisory Council interpreted its terms of reference more broadly than that and its 'plan of action' included support for both pure and applied research. In the first year of operation, £25 000 were set aside for disbursement by the Privy Council Committee on the advice of its Advisory Council and £40 000 were promised for the second year.

It soon became evident that administrative location within the Board of Education was inappropriate for science policy funding and co-ordination machinery and on 1 December 1916 Lord Crewe, the Lord President of the Council, announced the creation of a separate Department of State to be responsible for the government's scientific and industrial affairs. The Secretary of this new Department of Scientific and Industrial Research (DSIR) reported directly to the Lord President of the Council, who was assisted in his duties by a Committee of the Privy Council and advised by an Advisory Council as before. For its first five years of operation a budget of £1 million was promised to the DSIR. The Devonshire Commission's recommendation for a Ministry of Science and Education and the establishment of a Council representing the scientific knowledge of the nation had been partly followed.

The Lord President of the Council was not a Minister for Science but he was a minister without line department responsibilities. It was acknowledged that only scientists could judge the potential fruitfulness of projects submitted to the government for support, as the Devonshire Commission had suggested, and the Advisory Council performed this role.

Thus, from the very beginning of the government's serious involvement in scientific affairs, science was treated differently from the other state activities carried out by the administrative departments. The establishment of the DSIR was, therefore, a further milestone in the development of the concept of a national science policy. Central features of the DSIR's management and organization can be located in the science policy machineries of many countries today. These include a Minister without Portfolio as political head, a Council of Ministers or senior civil servants to assist the Minister without Portfolio in the execution of his duties, an Advisory Council of scientists and industrialists to advise the Minister and his Committee on their responsibilities and, finally, a Department of State like the DSIR, under the control of a Permanent Head, who is usually an eminent scientist in his own right, with responsibility for funding and performance of long-term and medium-term research.

The DSIR was not an administrative department like the War Department or the Colonial Office, however. The end product of its activities was research, an intangible commodity that could be of long-term benefit to other departments and to the nation as a whole. The administrative departments did have some interest in scientific research but their research needs were generally short-term and specific to their operations. The Admiralty and the War Department, for example, were supporting research in their own small research establishments when the DSIR was set up. But the Board of Trade, for example, would not at that time be particularly concerned to increase the volume of industrial research, nor would the Board of Education concern itself unduly with the output of research scientists from the universities. No Department of State was particularly concerned with pure research upon which, it was widely believed, the nation's long-term industrial prosperity depended, and research that did not fit into the areas of responsibility of the Departments of State, too, was neglected. The DSIR was intended to fill the gaps, to correct the bias towards short-term specific research that was prevalent in the other departments. There was no suggestion in government at the time that it would carry out or sponsor all governmental research so in this it differed from Alexander Strange's proposed Ministry in which all that was done in the nation's name to advance science was to be brought under one Minister.

The DSIR was the most significant and lasting organizational inno-

vation in British government to emerge from the First World War and
it set the pattern for similar developments in other parts of the Empire.
Canada's National Research Council, Australia's Commonwealth
Scientific and Industrial Research Organization, New Zealand's Department of Scientific and Industrial Research and the Councils for
Scientific and Industrial Research in South Africa, India and Ceylon
were all inspired by the example set in the mother country in 1915. In
Britain, a Committee of Inquiry into the Machinery of Government set
up in 1917 under the chairmanship of Lord Haldane confirmed the
arrangements that had been made during the War for the management of the government's scientific and industrial research effort. In
the Report of this Committee the so-called Haldane Doctrine for the
organization of government science was promulgated. This doctrine is
neatly encapsulated in the following passage:

> As regards the methods to be adopted for conducting inquiry and research
> in any branch of knowledge so far as it is determined that the work should
> be carried out under the supervision of a general organization and not
> under that of an administrative department, we think that a form of
> organization on the lines already laid down for the Department of Scienti-
> fic and Industrial Research will prove most suitable.[18]

According to the Committee, one of the advantages of such an organization for research of a general nature — research that is not geared to
the needs of a specific administrative department — was that it placed
responsibility to the government in the hands of a Minister who was, in
normal times, free from any serious pressure of administrative duties
and immune from any suspicion of being biased against the application of research results by administrative considerations. In accordance
with this principle, in 1919 governmental support for medical research
was placed under the control of a Medical Research Council, the
Secretary of which reported to the Lord President. Agricultural research joined the system in 1932, and the Nature Conservancy (now
the Natural Environment Research Council) in 1949.

Despite organizational innovations like the DSIR after the First
World War, there was no great upsurge of government support for
scientific research. War research, the great stimulus for all kinds of
other research, was inhibited by the military's innate conservatism, by
the repugnance towards war that the 'war to end war' had left in
people's minds, by economic depression, and not least by a feeling of
security that, in Britain, was reflected in the assumption as a military
policy guideline of a rolling ten-year period of peace, an assumption
that lasted until the end of the 1920s. Between the wars the DSIR co-
ordinated and expanded the government research effort, both civil
and military, but towards the end of the 1920s, with the growing
scientific activity of the three armed services, its co-ordinating role

began to wane. Its role as a research performer developed slowly but steadily after 1918 when it acquired the National Physical Laboratory, and in subsequent years fourteen other major laboratories were added. It administered the Industrial Research Association scheme, a form of government intervention that had begun during the First World War, and it continued to distribute research funds to universities until 1964.

The 1930s saw the emergence of another science policy 'theorist', J.D. Bernal, Professor of Crystallography at the University of London. Bernal was a leading member of the Marxist-oriented Association of Scientific Workers and one of the principal characters in the freedom/planning debate referred to in Chapter 3. In *The Social Function of Science*, published in 1939, Bernal surveyed the British research system of the time and found it to be inadequately funded, uncoordinated, poorly organized and insufficiently attuned to the needs of British society. Though his diagnosis was correct in many ways, he offered little in the way of a cure. While he clearly wished to influence the direction of science in Britain, he was rather short on the machinery with which this could be done. All he could say was:

> It should be possible to have a flexible and rational system of administration which would secure economic efficiency and at the same time preserve and encourage the characteristics of national and local cultures. Anarchic private interests and stupid bureaucracy can destroy most of the potential value from technical advances.

Shortly after the publication of *The Social Function of Science*, the Second World War broke out and science policy in Britain and elsewhere took a great leap forward.

The role of science and scientists in this war is well documented and it is only necessary to comment on the developments in science policy that took place first in the United Kingdom and then in the United States as a result of the conflict. Rose and Rose point to three themes of general importance that have emerged from the close relationship that developed between science and government in this truly scientific war. First, there was the rapid increase in war research, some of it concealed even from parliamentary scrutiny; second, there was the growth of centralized bodies for making decisions about national scientific and technological developments; and third, there was public recognition of the need to plan to retain these bodies as permanent features of the research system after the war. The second and third trends are illustrated by experience in Britain and the United States.

In Britain, a number of separate scientific advisory and executive bodies came into existence immediately before the outbreak of the war. In 1934, a Committee for the Scientific Study of Air Defence, under the chairmanship of Sir Henry Tizard, was set up by the Air Ministry when it became patently obvious that Britain was virtually defenceless against air attack. This Committee was soon to provide the mechanism by which radar, Britain's saviour in the first year of the war, could be developed. Shortly after its establishment, a memorandum was submitted to the Committee by R.A. Watson-Watt, Director of the National Physical Laboratory, which suggested that as a result of discussions that had taken place between him and his assistant he thought that it might be possible to detect enemy aircraft by using radio waves. Within a month, a successful demonstration convinced the Committee that work on radio-detection should be supported immediately and the technique was developed in the government's own research establishments. The Home Chain radar network was installed by the outbreak of war as a result of these initiatives.

In 1940, an overall Scientific Advisory Committee to the War Cabinet was created, comprising the Lord President of the Council and Secretaries of the DSIR, the Medical Research Council and the Royal Society, to advise the Lord President of the Council on any matter referred to the Council, to advise government departments when so requested on the selection of individuals for particular lines of scientific enquiry or for membership of committees on which scientists were required and, finally, to bring to the notice of the Lord President promising new scientific or technical developments which might be important to the war effort. This Committee was rather ineffective. Churchill preferred to seek the advice of F.A. Lindemann, his own personal adviser (who was opposed to the radar decision at the time), and he tended, therefore, to ignore or short-circuit the Committee.

In 1939, the Ministry of Supply, a Ministry that was closely associated with Defence, acquired a scientific advisory committee on scientific research and technical development and, by the end of the war, had three full-time scientific advisers. Special committees were also set up from time to time, the most famous of which was the so-called MAUD Committee. In March 1940, two 'alien' nuclear physicists working at the University of Birmingham, Rudolf Peierls and Otto Frisch, sent to Sir Henry Tizard a memorandum that outlined the probable effects of a 'Radioactive Superbomb' which could be constructed by inducing a chain reaction in the 235 isotope of uranium. The predictions of the Frisch-Peierls Memorandum, which were based on theoretical calculations derived from work already done in Germany, France and Italy, were remarkable. They estimated that the yield from a critical mass of U_{235} would be equivalent to 1000 tons of dynamite, that the temperatures achieved would be comparable to those found inside the sun, that radioactive contamination would be a serious hazard for some time after the explosion and that there would be radioactive fallout downwind of the explosion. Frisch and Peierls considered that since the German scientists Hahn and Meitner had first split the atom, it was reasonable to suppose that a full scale nuclear research program would, by then, be in operation in Germany. They warned that if Germany acquired the superbomb the war would be lost. In response to the Memorandum Tizard immediately set up a committee chaired by Sir George Thomson, Professor of Physics at the University of London and one of the few experimental scientists then working on the problem of nuclear fission in Britain. This was the MAUD Committee which was instrumental in getting Britain's atomic bomb project under way. But the age of 'big science' had dawned, and wartime Britain was unable to support a project of the magnitude required to develop the atomic bomb with great speed. Scientific and technological leadership was about to pass to the United States, a country that could, at this time, afford the millions of dollars necessary to develop the bomb.

Though Rose and Rose have suggested that one of the themes that emerges from a study of Britain's wartime research effort is the growth of centralized bodies for scientific decision-making,[19] in fact there was relatively little centralized decision-making or co-ordination of research during the war. Though the inner circles of the government had been penetrated by Professor F.A. Lindemann, who acted as chief scientific adviser to Churchill and the War Cabinet, his position owed more to his personal friendship with Churchill than it did to Churchill's or the government's high valuation of the scientific role in government. In general, science policy in Britain during the war developed in accordance with the principle that experts should be on tap but not on top. Nevertheless the science policy apparatus that was in place by 1940 worked well in two vital instances — radar and atomic energy.

In the post-war years, the scientific advisory council became a permanent feature of British science policy. In 1947, the Advisory Council for Scientific Policy (ACSP) replaced the Scientific Advisory Committee to the War Cabinet, and a Defence Research Policy Committee was also set up. In an attempt to achieve co-ordination, both Committees were under the chairmanship of Sir Henry Tizard, now back in favour after the Labour Party's victory in the 1945 elections, and this arrangement lasted until 1952, when Tizard retired. But apart from the introduction of advisory councils and a new agency to develop the British atomic energy program (the United Kingdom Atomic Energy Authority [UKAEA]), the British science policy apparatus remained essentially the same as it did before the war. Major changes did not take place until 1964. These issues will be discussed in Chapter 5.

Science policy in the United States: early developments

The history of the science–government relationship in the United States, right up to the beginning of the First World War, was one of mutual aloofness. On the one hand, scientists feared government inter-ference in their activities and were suspicious of any financial arrange-ments that might jeopardize their independence. The Federal Govern-ment, on the other hand, steeped in the liberal-democratic tradition, considered it improper for a government to intervene in an activity that should be controlled by market forces.

The National Academy of Sciences was founded in 1863 by an Act of Congress for the purpose of providing scientific advice to government agencies that requested it. The departments and agencies rarely sought the advice of the Academy, however, and although membership of the Academy became one of the greatest honours that could be bestowed upon an American scientist, the institution itself acquired little power. Some twenty years after the foundation of the Academy, an editorial in *Science*, the journal of the American Association for the Advancement of Science, commented:

> The liberal spirit which animates both Congress and the executive depart-ments in their dealings with scientific affairs is very apt to lead them into the support of scientific enterprises without any sufficient consideration of the conditions of success, and of efficient and economical administration, and a careful consideration of each proposed undertaking by a committee of experts is what is wanted to insure the adoption of the best methods.[20]

This was a call for the adoption of some means of evaluating scientists' proposals, preferably by the use of an advisory committee of scientific experts, a mechanism that is now familiar to us. At the time, perhaps stimulated by Alexander Strange's pronouncements on the subject in England, there was much discussion in the United States about the establishment of a Department of Science. The scientists and the bureaucracy were divided on the issue, and in 1884 Congress ap-pointed the Allison Commission to survey and study the procedures of handling science in other countries and to recommend methods of co-ordinating the scientific areas of the government agencies that were then engaged in research. The Allison Commission sought the assis-tance of the National Academy of Sciences, which established a com-mittee to formulate its advice. This committee suggested that the time was ripe to create a branch of the government to be responsible for the direction and control of all its scientific work. If this could not be achieved, then the committee recommended that a co-ordinating com-mittee be set up. Neither a department nor a co-ordinating committee was, in the end, set up, and government support for science and the influence of scientists in government remained small.

The First World War provided the opportunity for re-assessment of the science–government relationship in the United States. During the conflict an advisory group, the National Advisory Committee for Aeronautics (NACA), was established. This Committee, which re-ported to the President, was composed of government and non-government experts acquainted with the needs of aeronautical science and it was set up because 'every first-class nation in the world except the United States had an advisory committee in aeronautics.'[21] Apart from the establishment of NACA, however, there was little change in science policy in the United States as a result of the First World War.

During the Great Depression, as a result of the joint efforts of leaders of the scientific community within the Academy of Sciences and of the Secretary for Agriculture, a Science Advisory Board was created by Executive Order in order to advise the President. It operated through the machinery, and under the jurisdiction, of the National Academy of Sciences and was composed of nine members, the majority of whom were academics who were anxious to gain government support for pure research in the universities. Shortly after its establishment, it requested US$75 million for such research, with control over the administration and allocation of the funds to be in the hands of the scientific community itself. The government, predictably, refused the request and the Board was wound up in 1935, with a pile of reports and a few changes in the scientific programs of some government agencies to its credit.

The next move towards the development of a science policy machinery came in June 1940. Knowing that the war then raging in Europe was a truly scientific conflict, the leaders of the scientific community began to lobby for the creation of a government agency that would be able to mobilize American science and scientists for the country's inevitable entry into the war. As a result, the National Defense Research Committee (NDRC) was created by President Roosevelt under the chairmanship of Dr Vannevar Bush, former Dean of Engineering at the Massachusetts Institute of Technology and President of the Carnegie Institution in Washington. Members of the committee were drawn from the universities, government departments and agencies, private foundations and industry. Its mandate included the correlation and support of scientific research on the mechanisms and devices of warfare, with the exception of those relating to war in the air. The committee was directed to aid and supplement the experimental and research activities of the War Department and the Department of the Navy, and was authorized to conduct research into the creation and improvement of instrumentalities, methods and materials of warfare. It was authorized to use the laboratories and equipment of the government's own scientific establishments insofar as these were available, and most importantly, it was authorized to offer contracts to individuals, to educational or scientific institutions (including the National Academy of Sciences), and to industry for studies and experimental investigations to be carried out.

After about six months of operation, Bush urged upon the government some changes in the NDRC's operation. Asked by the Bureau of the Budget to prepare a report on the optimal organization of defence research, Bush made three points in his reply. First, he objected to the emphasis on research alone within the NDRC. He was not convinced that development work should be handed over to the military because he had little faith in their ability to carry it through. Research groups, he argued, should be given the responsibility to carry projects right through to prototype stage. Second, there was little co-ordination of the research activities of the NDRC, the armed services and the NACA. Third, military medicine, an important field of endeavour, was being neglected under the arrangements then in force. The outcome of this report was the establishment by Executive Order of the Office for Scientific Research and Development (OSRD), an operational agency with two advisory committees, the NDRC and a Committee on Medical Research.

The creation of the powerful OSRD, under the directorship of Dr Bush, was a landmark in United States science policy. It brought the American scientific community into the war effort but, through the contract mechanism, left the scientists much of the freedom they thought was essential for creative work. The contractor simply agreed to carry out studies or experimental work in connection with a given problem and to make a report to the customer (OSRD). No attempt was made to dictate the manner of the investigations, nor were the contractors compelled in any way to take on any tasks. Through the OSRD, for the first time in American history, substantial government funding was being channelled to the universities, either directly or by the establishment of associated laboratories. The most famous of these laboratories were the Radiation Laboratory (RadLab) at the Massachusetts Institute of Technology, which carried out vital work on radar, the Metallurgical Laboratory at the University of Chicago, where the first atomic pile was assembled and a sustained chain reaction demonstrated in 1942, and the Los Alamos Atomic Weapons Laboratory associated with the University of California, where the first atomic weapons were designed and manufactured under the direction of J.R. Oppenheimer.

There is no doubt that the OSRD was the most important and most powerful science policy agency created in the United States during the Second World War, and it played a leading role in getting the American atomic bomb project under way. Towards the end of the war, President Roosevelt asked Dr Bush to recommend how the OSRD's experience in wartime could be utilized after the war. Bush was asked to consider the following four questions:[22]

- the extent to which wartime successes could be applied in peacetime for the benefit of the nation as a whole;
- the organizational structure that might be erected in order to con-

tinue the work on medical research that had proved to be so successful during the war;

- the role of the government in fostering research activities by public
  and private organizations;
- the ways in which scientific talent might be developed in the United
  States in order to assure the dominance of American science in the
  future.

Bush sought submissions from interested bodies, and in July 1945
submitted his report, entitled *Science, the Endless Frontier*, to President Truman. Among the recommendations were that:

- the government should provide financial support for basic medical
  research in the medical schools and universities. (The Committee for
  Medical Research within OSRD was doing this at the time, but there
  was no guarantee that financial support would continue after the
  war);
- there should be more, and more adequate, military research in
  peacetime through a civilian-controlled organization with close
  liaison with the Army and the Navy but with direct support from
  Congress;
- the government should promote industrial research by increasing
  the flow of new scientific knowledge through the support of basic
  (pure) research, by aiding in the development of scientific talent
  and, finally, by providing suitable incentives to industry to conduct
  research;
- a permanent Science Advisory Board, composed of disinterested
  scientists having no connection with the affairs of any government
  agency, should be established to advise the executive and legislative
  branches of government on the policies and budgets of those government agencies that were engaged in scientific research. The Board
  would also serve as a co-ordinating body;
- undergraduate scholarships and graduate fellowships should be provided in reasonable number;
- in order to put these recommendations into effect a National Research Foundation should be established, receiving its funds from
  Congress to (a) develop and promote a national policy for scientific
  research and scientific education, (b) support basic research in non-
  profit organizations, (c) develop scientific talent by means of
  scholarships and fellowships, and (d) by contract and otherwise,
  support long-range research on military matters.

The proposal for a National Research Foundation was the most important of Bush's recommendations. He suggested that the agency
should:

- be composed of civil service personnel of broad interests and experience, having an understanding of the peculiarities of scientific research and scientific education;
- have stability of funds so that long-range programs might be undertaken;
- recognize that freedom of inquiry must be preserved and leave internal control of policy, personnel and method and scope of research to the institutions in which they are carried on;
- report to the President and, through him, to Congress.

Responsibility for the agency was to be placed in the hands of nine members, who should be persons not otherwise connected with the government and not representative of any special interest, and these should be selected by the President. The members would elect their own chairman and, importantly, appoint the Executive Director of the Foundation.

Bush's report was a classic 'policy for science' document. It was directed almost entirely towards the support of pure research by the government, and it gave the scientific community almost complete control over the allocation of government funds for pure research. The only element of co-ordination in the report is directed towards the government agencies which, it assumed, conducted applied research only, and this co-ordination would be achieved by the disinterested scientists in any case. Pure research could not be co-ordinated or controlled. In Bush's words:

> The government is peculiarly fitted to perform certain functions, such as the co-ordination and support of broad programs on problems of great national importance. But we must proceed with caution in carrying over the methods which work in wartime to the very different conditions of peace. We must remove the rigid controls which we have had to impose and recover freedom of inquiry and that healthy competitive scientific spirit so necessary for expansion of the frontiers of scientific knowledge.[23]

Bush considered that the government would confer such freedoms on the scientific community because of the thinking at the time about the relationship between pure science and technological invention and innovation. This thinking is summed up in the following passage from his report:

> Basic [pure] research leads to new knowledge. It provides scientific capital. It creates the fund from which the practical applications of knowledge must be drawn. New products and new processes are founded on new principles and new conceptions which, in turn, are painstakingly developed by research in the purest realms of science.[24]

This excerpt illustrates what is now known as the linear model of innovation, a view of innovation that assumes it to be the product of prior research undertaken without thought of any practical application. (The linear model of innovation is discussed in Chapter 2.) From Bush's perspective, it was reasonable to assume that the government would accept a premise that had been so spectacularly demonstrated by the atomic bomb project, although no systematic research had been done into the relationship between investment in pure science and pay-off in terms of technological innovations.

However, President Truman doubted the wisdom of giving the scientific community almost complete control over the distribution of government funds to universities and other private foundations and he vetoed the first National Science Foundation Bill after it had passed through the Legislature in 1947. Three years were to elapse before it could be presented again. The Army, Navy, and Health, Education and Welfare (HEW) Departments took advantage of the delay. The military especially now recognized the importance of scientific research and had the choice of either building up the research capacity of its own establishments or forging a close relationship with the universities, or both. The former course was not likely, since many of the best scientists available after the war would not choose a civil service career, so it was decided to conduct as much classified research as possible intramurally and to contract much of the basic research to the universities.

In 1946 the Office of Naval Research (ONR) was created to 'plan, foster and encourage scientific research and to provide within the Department of the Navy a single Office which, by contract or otherwise, shall be able to obtain, co-ordinate, and make available to all bureaux the necessary services for conducting specialized and imaginative research.'[25] At the time of its establishment, the Navy was already supporting contracts worth more than US$22 million in university and private industrial laboratories. The Office established an excellent working relationship with the universities and convinced them that it was quite appropriate for them to accept contracts from the military, even in peacetime. It was not only research that might be relevant to Navy needs that was funded, however. As Penick et al. have demonstrated, the men who made the ONR a success believed that the primary aim of much of the research that it funded should be free rather than directed—that is, it should explore and understand the laws of nature and not necessarily be aimed at solving practical problems.[26] Moreover, the naval authorities believed that most of the basic research carried out under the ONR's auspices should be published in the normal way. Taken together, these 'conditions' allayed the fears of university administrations and the scientific community, and the Office began to take on the role that was envisaged for the National Science Foundation.

The United States Public Health Service, later to become the Department of Health, Education and Welfare, also took steps to fill the gap left by the National Science Foundation's failure to get off the ground. Its National Institutes of Health, established in 1930 and generously funded by the OSRD during the war, became the focus of government support for medical research. By 1951, their expenditure on intramural research and research contracted to university medical schools had reached US$30 million. The Atomic Energy Commission was established in July 1946, and this too forged quite close links with the universities by contracting research work to them and by building up the university-associated laboratories that it had inherited from the Manhattan Project. These were re-designated National Laboratories, and some, for example the Oak Ridge, Lawrence, Argonne and Brookhaven National Laboratories, have become very famous as centres of scientific research, both basic and applied.

Thus, science policy in the United States in the immediate post-war years was dictated by individual administrative departments rather than by a specific plan of action by the Federal Government. As Lambright has noted in his excellent review of the period, 'whatever was said or done at the highest levels of government about science, the reality of policy was occurring at the administrative level,'[27] and this was because the government had delegated great responsibility to the mission agencies for recognizing scientific opportunities, creating new scientific initiatives and anticipating the nation's long-range needs for research and development. But there was little co-ordination of the scientific activities of these departments, despite the recommendations of the Steelman Report which had been submitted to President Truman in 1947. In 1946 Truman, encouraged by a group within the Executive Office who objected to the whole tenor of Bush's *Science, the Endless Frontier*, established a President's Scientific Research Board under the chairmanship of John R. Steelman, one of the President's assistants. The four-volume report, entitled *Science and Public Policy*, made numerous recommendations, some of which supported Bush's recommendations and some which did not. The most important of these were:

- that annual expenditure on R & D should reach the level of 1 per cent of the GNP by 1957 (it was, at the time, about 0.5 per cent);
- that heavier emphasis be placed on basic and medical research, and that expenditures on basic research be quadrupled and on medical research tripled by 1957;
- that the Federal Government support basic research in universities and non-profit organizations to the extent of US$250 million by 1957;
- that a National Science Foundation be established with a Director appointed by the President and reporting to him, with a part-time Board of eminent scientists and educators, half to be drawn from outside the Federal Government and half from within it;

- that a federal committee be established, composed of the directors of the principal federal research establishments, to assist in the co-ordination and development of the government's own R & D programs;
- that a number of co-ordinating centres be established within the Executive Branch of the Government to formulate policy. These co-ordinating centres would comprise (a) an Interdepartmental Committee for Scientific Research, (b) a unit within the Bureau of Budget for reviewing federal scientific R & D programs, and (c) a science liaison officer in the White House.

Acting on the Steelman Report, Truman established an Inter-departmental Committee on Scientific Research and Development to study and report upon current policies and federal administrative practices relating to federal support for research. This Committee, which comprised representatives of federal agencies, failed as a device for co-ordination of agency research programs.

When the National Science Foundation was eventually established, it was hoped that this would monitor and co-ordinate scientific research and development. The Bill that brought the National Science Foundation into legal existence in July 1950 authorized it to develop and encourage a national science policy to promote basic research and education in the sciences and to evaluate scientific research programs undertaken by the agencies of the Federal Government. The funding of medical and defence research was excluded from its terms of reference and in the first few years of its operations its budgets were small compared with the research expenditures of the big departments and agencies. Not surprisingly, it failed to act as a co-ordinating agency of any sort. In the words of Don K. Price, 'a new agency with a budget of a few million dollars does not realistically undertake to give orders to the great departments with budgets of billions, no matter what the law may say.'[28] In 1954, an Executive Order of the President attempted to reaffirm the Foundation's monitoring and co-ordinating roles. Its terms of reference as set out in the Order directed the Foundation to:

- recommend to the President policies which would strengthen the national scientific effort and help define the responsibilities of the Federal Government in the conduct and support of pure research;
- make comprehensive studies and recommendations regarding the nation's research effort and its resources for scientific activities, including facilities and scientific personnel, and its foreseeable scientific needs, with particular attention to the extent of the Federal Government's activities and the resulting effects upon trained personnel;

- review the Federal Government's scientific research programs and activities in order to formulate methods for strengthening the administration of such programs and activities by the responsible agencies and to study areas of basic research where gaps or undesirable overlapping of support may exist and make recommendations to the heads of agencies concerning the support given to basic research;

- provide support for general purpose basic research through grants and contracts. The conduct and support by other federal agencies of pure research in areas closely related to their missions were recognized as important and desirable, especially in response to current national needs, and would continue;

- consult with educational institutions, heads of federal agencies and the Commissioner of Education of the Department of Health, Education and Welfare, and study the effects upon educational institutions of federal policies and the administration of contracts and grants for scientific R & D;

- recommend policies and procedures which would promote the attainment of general national research objectives and realization of the research needs of federal agencies, while safeguarding the strength and independence of the nation's institutions of learning.[29]

The head of each federal agency engaged in scientific research was to make certain that effective executive, organizational and fiscal practices existed to ensure (i) that the Foundation be consulted on policies concerning the support of basic research; (ii) that approved scientific research programs conducted by the agency would be continually reviewed to preserve priorities in research efforts and make adjustments for changing conditions without imposing unnecessary added burdens on budgetary and other resources; (iii) that applied R & D should be undertaken with sufficient consideration of the underlying basic research and other factors such as relative urgency, project costs and availability of manpower and facilities; and (iv) that, subject to considerations of security and applicable law, reports on the nature and progress of research projects would be adequately disseminated within the Federal Government as an aid to efficiency and economy of the overall federal scientific research program. But the conditions that had led to the Foundation's failure to act as a monitoring and co-ordinating body were no different in 1954 from those that prevailed in 1951, and apart from a brief period during the Nixon Administration, when its Director became the President's Science Advisor, the Foundation has never played a major role as a policy-making or co-ordinating agency.

Since defence research was excluded from the National Science Foundation's terms of reference, and the President was concerned that the arrangements existing in 1950 for liaison between departments and agencies concerned with R & D were inadequate, a Science Advisory Committee of scientists and industrialists was set up within the Office of Defense Mobilization. Its terms of reference were to:

• provide independent advice on scientific matters, especially as regards the objectives and interrelations of the several federal agencies engaged in research of defence significance;
• advise on progress being made in dealing with current scientific research problems of defence significance and concerning defence research matters which needed greater attention or emphasis; advise on plans and methods for implementing scientific effort for defence;
• transmit the views of the nation's scientific community on R & D matters of national defence significance.

In 1957, as a result of Russian successes in space, the Science Advisory Committee was removed from the Office of Defense Mobilization to the White House and reconstituted as the President's Science Advisory Committee (PSAC), which reported to the President and to the newly appointed President's Special Assistant for Science and Technology, Dr James R. Killian, Jr. Two years later, on PSAC's recommendation, a Federal Council for Science and Technology (FCST) was created to formulate and co-ordinate the government's research policy and planning. Like Truman's Interdepartmental Committee on Science and Technology, established almost a decade before, the FCST was composed of delegates from government departments and agencies, and it too proved to be a disappointment.

Subsequent science policy developments in the United States will be dealt with in more detail in Chapter 4. The Second World War had demonstrated conclusively the link between knowledge and power, and as a result science became a major instrument of United States government policy. The instrument was shaped and developed by the departments and agencies, not by the Executive Office. By 1957, impelled by the Cold War, expenditure on scientific research had risen far beyond that suggested in the Steelman Report, and after 1957, as a result of the Sputnik scare, it rose even faster than before. A science policy machinery was established which gave representatives of the scientific community access to the highest levels of the government, and the existence of the National Science Foundation ensured that, even if the mission departments and agencies lost their enthusiasm for pure research, this research would not be entirely neglected by the government.

The European scene

The first attempts to apply the lessons of the wartime organization of science to the problems of peacetime economic reconstruction and the maintenance of a credible defence posture were made in Britain. In mainland Europe, international organizations played an important role in shaping the science policy machineries that were eventually established in most European countries. The Organization for Economic Co-operation and Development (OECD), based in Paris, was especially influential. In advocating particular forms of science policy machinery and in providing much of the empirical knowledge about such things as research and development expenditures and attitudes towards research in member countries, the Organization provided governments with some insights into how they might organize their own national research and development efforts.

In 1949 the OECD's predecessor, the Organization for European Economic Co-operation (OEEC), issued a report in which member countries were urged to run their industries on more scientific lines. Operational research, a scientific approach to management and the measurement of productivity were all advocated, and member countries were asked to recognize the importance of R & D for their industrial competitiveness and economic growth in the coming decades. Ten years later, the OECD realized that the R & D of member European countries seemed to be relatively ineffective when compared with that of the United States. As a result it commissioned Mr Dana Wilgress, former Canadian Ambassador to the OEEC, to discuss with government scientists and industrial leaders the measures taken or planned to acquaint the decision-makers in government and private industry with the importance of science and technology for their economies and their survival, and to propose national and international measures that would increase technical resources and encourage co-operation with a view to using their resources more effectively.

The Wilgress Report was the first of many OECD reports dealing with national science and technology policies. It concluded that member countries had failed to appreciate the importance of science and technology. They were loath to scrap traditional methods and reluctant to adapt their educational systems to the needs of science and technology. For our purposes, the two most important recommendations were that member countries should draw up science policies to take into account the need for a suitable balance between pure and applied research, and that they should concentrate on science and technology as the bases for economic growth. Importantly, Wilgress recommended that scientific affairs be consolidated within the OECD by the establishment of a high-level science policy advisory group. This advisory group was established in 1960 under the chairmanship of Pierre Piganiol, Chief Scientist for the French Government. In 1961 it produced a report entitled *Science and the Policy of Governments*, a document that provided the focal point for discussions about science policy as we now know it. This report recognized the need to develop criteria by which resources might be allocated to the various types of scientific research activities. It recognized that the environment for the conduct of research needed to be carefully considered so that scientific and technological resources might be applied effectively to national problems and, importantly, it separated science policy into two elements, 'science for policy' and 'policy for science'.

> The term 'science policy' is ambiguous. It too often connotes only a policy limited to the needs of science *per se*, and excludes the effects of science and technology on the full spectrum of national policies in such disparate fields as agriculture and industry, defence, education and domestic and foreign political affairs. Maximum exploitation of scientific opportunities requires programs that combine concern for the growth of science itself and provision for the rapid deliberate application of its fruits to human welfare. That is the substance of science policy in the full sense, as denoting consideration of the interactions of science with policy in all fields.[30]

The report's recommendations included the following:

> each government should consider setting up a central mechanism to discuss science policy in the broad sense defined; the OECD should convene a meeting of ministers with responsibility for science so that matters of mutual concern might be discussed.

The first ministerial meeting took place in 1963 at a time when only three of the twenty-two member countries had special ministers for or of science. The other countries were represented mainly by Ministers of Education. As we shall see in Chapters 4 to 8, this is no longer the case. Ministerial meetings have become a regular feature of OECD science policy activities, and special reports, which review general trends in science policy and make recommendations for its future direction, are often commissioned for the ministers' consideration. At the second ministerial meeting, held in 1966, it was agreed that standardized R & D statistics would be collected so that the ministers could compare their own efforts with those of their fellow members. After this, a program of reviews of national science policies was begun and took the following form:

- a background paper on the research system of the country under review was prepared by OECD in consultation with relevant national authorities;

- examiners used this report as their starting point for an investigation of the country's science policy system;

- 'confrontation' meetings between appropriate senior officials, the OECD examiners and science policy decision-makers from other countries allowed strengths and weaknesses of the national science policy system to be exposed and commented upon;
- a report was published.

These reports are important source documents and provide guidelines for countries whose science policies are evolving and changing.

The OECD has played a significant role in institutionalizing the concept of science policy as defined here. It has pioneered the collection of R & D statistics, providing the basic information by which member countries can evaluate their own systems. The OECD has its own definite views on what science policy machinery should entail, and its recommendations for Australia are typical:

- a Minister for Science without responsibility for a line department, in other words, a Minister for Science Policy;
- a co-ordinating Ministerial Committee chaired by the Minister for Science;
- a powerful advisory council assisting the Ministerial Committee, reporting to the Prime Minister through the Minister for Science.

This is the so-called concerted action model of science policy machinery which will be discussed in Chapter 3. The OECD's broad evaluation of trends in science policy will be considered in the last chapter of this book.

CHAPTER 2

Intervention and innovation

Technological innovation and the economy

Towards the end of the Second World War, when the scientists were at the height of their influence as a result of their wartime achievements, President Roosevelt asked Vannevar Bush, the Director of the Office of Scientific Research and Development, to prepare a report that would spell out the advantages to be gained by applying the information, techniques and research experience learned during the war to post-war economic and social problems. Dr Bush was aware of the fact that the scientists' accomplishments in the First World War led to no great increase in government funding for either pure or applied research, and he was concerned to ensure that this did not happen again. Though the scale of scientific effort and the decisiveness of the scientists' contributions to victory in the Second World War made it unlikely that the scientific community would return to its pre-war state of relative impoverishment, Bush and other leaders of the American scientific community felt that a positive effort was needed to make sure that the close wartime relationship that had developed between science and government would continue. Only through such involvement could the massive resources that pure science now needed become available.

In a nation that placed great store on inventiveness, ingenuity and practical utility, it was important for Bush to emphasize the practical benefits that would surely arise from investment in pure research — that long-neglected area of American scientific endeavour. This is precisely what he did in *Science, the Endless Frontier*, which, as has been pointed out, is a classic 'policy for science' document. He stressed the indispensability of pure research to the achievement of good health, economic prosperity and national security. Advances in pure research would lead to more jobs, higher wages, shorter working hours, more abundant crops, more leisure, and the prevention and cure of disease. Without scientific progress, the nation's health would deteriorate, there would be no improvement in the standard of living or unemployment and the constant risk of defeat by aggressors would remain.

Significantly, Bush emphasized the point that the financial support of pure research is a proper concern of government, since 'there are areas of science in which the public interest is acute but which are likely to be cultivated inadequately if left without more support than will come from private sources.'[1] Pure research, which Bush considered to be the pacemaker of social and economic progress, was one such area. 'New products and processes are founded', he said, 'on new principles and new conceptions which, in turn, are developed by research in the purest realms of science.'[2] The policy implication of this view of pure research is clear. The simplest and most effective way of strengthening applied research in industry and government agencies was for the government to provide financial support for pure research in the universities, medical schools and non-profit institutions. Like Babbage in Britain more than one hundred years before, Bush was advocating government intervention in the scientific research system for economic reasons.

In giving pure research a key role in the innovative process, Bush was propounding what we now call the linear model of innovation. In this model, technological innovation, which for the time being may be defined as the commercial exploitation of a novel technical idea, is represented as a multi-stage process in which pure research is the first essential step. In 1945 this description of the innovative process was unquestioned and remained so until the mid-1960s, when science policy analysts carried out studies that threw some doubt on its accuracy. These studies will be discussed later on in this chapter. Bush used the model to persuade his government to 'intervene' in an activity that, as an element of a nation's cultural and intellectual heritage, would only receive a fraction of the funds it needed in the coming age of 'big science', and as an essential element of the innovative process would normally be regarded as the private sector's responsibility. The case had to be made by casting doubt on the private sector's willingness or ability to support pure research adequately.

Pure research did, of course, receive massive support in the United States after the Second World War, and this probably would have occurred even if Bush had never written his Report to the President. The Army and Navy recognized the need to keep the scientists on their side and heavily supported their more esoteric activities. The Department of Health, Education and Welfare sought and got control of medical research in post-war 'power play', and the Atomic Energy Commission's work received top priority in the Cold War atmosphere of United States foreign policy in the late 1940s and early 1950s. Only the much delayed National Science Foundation resulted from Bush's report, and this, because medical and defence research were removed from its sphere of operations, was but a shadow of the organization that he had envisaged. Departments and agencies in the United States, and similar bodies in other countries, supported pure research because they were convinced of its value, a conviction that rested on the evidence of the Second World War. It was when the impact of that evidence began to lessen and questions were being asked about the wisdom of government intervention on such a massive scale that the key economic arguments justifying government intervention in the entire research system, both basic and applied, were formulated. Before considering the theoretical arguments that justify government intervention, the terms should be defined.

At one time, it was customary to categorize scientific research as either pure or applied. Pure research was generally defined as research which resulted in general knowledge and understanding of the laws of nature, and it was pursued without regard to its immediate or even long-term practical utility. Applied research was defined as research oriented towards solving practical problems in industry or the government service. Results obtained by pure research workers in the universities were usually considered to be essential in solving these problems. As T.H. Huxley put it in 1882, 'What people call applied science is nothing but the application of pure science to particular classes of problems.'[3] These definitions were never satisfactory, but it was not until governments became major funders of scientific research and international comparisons became the order of the day that attempts were made to define it in a more systematic way. There is still no consensus on the matter. In this book, an adaptation of the OECD definitions will be used. In this classification, which is used by many government agencies in statistical compilations of their R & D expenditures, scientific research is divided into two broad categories—basic and applied. Basic research is further subdivided into pure research and strategic research. The precise definitions are shown in Table 2.1.[4]

The word technology is derived from the Greek *techne*, which means art, skill, craft, as well as a work of art, skill, craft. Strictly speaking, technology means the theory and principles of techne, just as psychology means the theory and principles of psyche, and it is often used (correctly) to denote a body of knowledge about techniques (practical methods of production). More often, however, the word technology is used incorrectly to describe the hardware that results from the application of technique. In this book, we shall use Mansfield's definition of technology as 'society's pool of knowledge regarding the industrial arts. It consists of knowledge of the principles of physical and social phenomena, knowledge regarding the application of these principles to production, and knowledge regarding day-to-day operations of production. Technological change is advance of technology and this may take place in a number of ways. Thus, new methods of producing existing products, new designs which result in

Table 2.1

BASIC RESEARCH

Original investigation with the primary aim of a more complete knowledge or understanding of the subject under study.

Pure Research

Basic research carried out without working for long-term economic or social benefits other than the advancement of knowledge and no positive efforts being made to apply the results to practical problems, or to transfer the results to sectors responsible for application.

Strategic Research

Basic research carried out with the expectation that it will provide a broad base of knowledge necessary as the background for the solution of recognized practical problems.

APPLIED OR TACTICAL RESEARCH

Original investigation undertaken in order to acquire new knowledge, and directed primarily towards a specific practical aim or objective such as determining possible uses for findings of basic research or solving a recognized problem.

EXPERIMENTAL DEVELOPMENT

Systematic work drawing on existing knowledge gained from research and/or practical experience that is directed to producing new materials, products and devices, to installing new processes, systems and services and to improving substantially those that are already produced or installed.

---

products with important new characteristics or new techniques of organization, marketing or management, can all represent a technological change.'[5]

Technological change involves invention, defined as a novel idea, sketch or model for a new or improved product, process or system. The first exploitation of a product, process or system of organization, either commercially or by a government, is an *innovation*, and this may, but need not, embody one or more inventions. A managerial or organizational change that leads to increased productivity, for example, would not involve invention but, according to this definition, is a technological innovation.

Until relatively recently, economists treated invention and technological change as an exogenous variable in models of the economic system; it was independent of any economic (market) forces, proceeding according to its own internal laws. Savings and capital accumulation were considered to be the principal agents of change in the economy, and economic growth could be understood in terms of the use of more and more inputs of capital and labour. In the 1920s, however, an American economist, Joseph Schumpeter, departed from this neo-classical tradition. Though he still accepted the neo-classical form of perfect competition as a suitable description of some types of markets in the short term and regarded innovation and technological change as exogenous factors, he suggested that in the long term competition based on innovation was the most important. This type of competition struck at the very survival of rival companies and not just at their output or profit margins.

Schumpeter therefore placed innovation at the very centre of his theory of economic development. His definition of innovation was broad, including the introduction of a new product or a new method of production, the utilization of new raw materials or the opening up of a new market or sector of the economy. He stressed the importance of entrepreneurs in bringing together new technical and socio-economic forces leading to innovation, arguing that although inventions and discoveries had been made throughout history, it was only when dynamic enterprising figures systematically applied new technical ideas in the production and marketing of goods and services that there was sustained economic growth. Importantly, Schumpeter made a sharp distinction between invention and innovation. Invention need not necessarily lead to innovation, he reasoned, and there could be innovation without technical invention, as when entrepreneurs combine existing elements in new ways.

So important was innovation to Schumpeter that he considered it to be the autonomous cause of repeated economic cycles which he called Kondratiev cycles, after the Russian economist who had earlier identified long waves of economic growth and recession. These long waves were the result of changing economic conditions for innovation and were not the result of a fluctuating rate of technical invention or other factors involved in the innovative process. Technical invention was, however, important in triggering them. The Kondratiev cycle theory has recently been revived in an attempt to understand the role of technological change in job creation and displacement.[6] Although technical invention was regarded as exogenous in his theory, and it was only one of a number of factors involved in innovation, Schumpeter gave it an importance that had been denied to it by his colleagues in the neo-classical tradition. He is an important pioneering figure in the study of the role of invention and innovation in economics.

After Schumpeter, and especially after the Second World War, economists focused much more sharply on the role of science and technology in the economic system. War had shown conclusively that science, technology and invention could respond to military needs and there seemed to be no reason in principle why they could not respond to non-military economic needs as well. Perhaps they had so responded in the past, and economists had not noticed their importance. Quantitative economic studies designed to throw some light on the relative importance of the various inputs into the economic system had shown that economic growth could not be adequately understood in terms of the two classical input factors of capital and labour alone. Continuing increases in gross domestic product per inhabitant could only be explained as learning to use existing inputs more effectively. With this realization came a renewed interest in technological change as a possible source of this 'third factor' of growth. Denison, in a study of United States economic growth, estimated that between half and three quarters was due to third factors and he concluded that advances in knowledge, both technical and managerial, were the most important part of the third factor. Most of the post-war economic growth in the United States depended, therefore, on the third factor, and technical invention, either as an exogenous variable or induced by economic factors, was, according to Denison, an important part of the third factor.[7]

There has been considerable criticism of the ways in which Denison and other economic growth modellers arrived at their precise percentage contributions of the various inputs to economic growth, but they did focus economists' attention on technological change, the diffusion of technology and the relationship between investment in scientific activities and output. In 1958, in a study of the diffusion of the hybrid corn innovation, Grilliches showed that diffusion of innovations could be explained in economic terms — on the basis of profit expectation as shaped by market size.[8] Some years later, with the work of Jacob Schmookler, the economists came full circle in their consideration of the role of technical change in economic growth. By examining patent statistics over a lengthy period in the agriculture, paper, petroleum and railroad industries, Schmookler came to the conclusion that not only could the diffusion of technological innovation be explained in terms of market forces, but the pattern of invention itself could be so explained.[9] Demand was the major determinant of variations in inventive effort in specific industries. In the railroad industry, for example, he found that increases in inventive activity, as measured by patents, lagged slightly behind increases in purchases of railroad equipment. Schmookler argued that variations in equipment purchases induced these variations in inventive activity. The data he gathered in all four industries indicated to him that inventors, corporate or otherwise, see increased equipment purchases in an industry as an indication of increased profitability of inventions in that industry and direct their inventive talents and resources accordingly. Technical change was not an exogenous variable after all; it was not independent of economic forces. Invention was controlled by demand considerations in the economy.

Schmookler acknowledged that there were limitations to the extent that demand could call up the appropriate inventions. Some inventions are impossible and others will seem impossible until the knowledge upon which they are based is uncovered. Clearly, the growth of scientific knowledge was important in Schmookler's scheme, since this would influence the characteristics of an invention; inventors, he argued, choose the most efficient means of achieving their ends and will select the area of science that will allow them to do this. Thus, the characteristics, but not the purpose, of an invention will be affected by this choice. Schmookler's view of invention is neatly summarized in the following quotation taken from his book. 'Invention,' he says, 'and in all possibility technical change itself, generally, is usually not apart from the normal processes of production and consumption, but a part of them. It expresses something not adventitious for a nation's life but an inherent part of it.'[10]

The market failure theory and its critics

Assuming Schmookler's analysis of the impact of economic demand on technical invention to be correct, the question arises as to whether an activity that is an inherent and vital part of national economies should be left entirely to the scientists, engineers and entrepreneurs of the private industrial sector. At the beginning of this chapter, it was noted that for military reasons alone it was inevitable that the United States and United Kingdom governments would continue to be involved in R & D. In the American case, this was a significant departure from the traditional science and technology policy. The British Government had, of course, been interventionist since 1915, when the Department of Scientific and Industrial Research was established, but its intervention was mainly inspired by military necessity, not economic advancement, and at the end of the emergency the level of government involvement dropped significantly. Neither in Britain nor the United States was this allowed to happen again, and there is hardly a government, and certainly none in the advanced industrial nations, that is now prepared to leave the production and utilization of scientific knowledge relevant to military and economic advancement entirely to the private sector. Many economists now accept the view that government intervention in the scientific research system is necessary if nations are to derive the greatest benefit from technological development.

The key papers setting out the justification for government intervention in R & D were written more than twenty years ago by Richard Nelson[11] and Kenneth Arrow[12]. They argued that, because of indivisibility, inappropriability and uncertainty, the private sector if left to itself would be expected to invest in research, especially basic research, at a socially sub-optimal level.

Indivisibility can be of two kinds. In conditions of atomistic competition, where many small companies compete in an industry in which social and economic benefits are likely to be increased by the rapid diffusion of technological change among all the producers, and cost effectiveness would preclude the individual producers from engaging in research on their own account, there is a situation of indivisibility. The industry as a whole may benefit from a centralized research effort and this could be provided either by government or by the industry itself. Farming is a good example of such an industry. The second type of indivisibility is where the R & D costs involved in bringing some modern high technologies to the market place are so high as to be beyond the financial capacity of even the largest private companies, which could not contemplate such ventures without government assistance. Civil supersonic aircraft development, fission and fusion nuclear reactors and large computer projects are examples of this type of indivisibility.

Inappropriability results from the fact that scientific research results are usually expensive to produce but relatively inexpensive to reproduce by commercial rivals. The patent laws give some protection, of course, but even in the event that an innovative firm exploits research successfully, it is unlikely that the firm will be able to appropriate to itself all of the benefits of its investment. Patents eventually run out and there may be applications of the knowledge produced that the innovating firm does not anticipate when the patent is filed. It is generally impossible to obtain patent protection for all future applications of a scientific discovery, so a discovery made in a firm's laboratories without any apparent application, and published in the open literature, may be exploited by commercial rivals. This is clearly a disincentive for firms to invest in research, especially pioneering, long-range research. In such circumstances, where the overall return on the long-term investment — in terms of such things as improved living standards and general social welfare — is greater than the return to a particular firm, there is a case for government intervention.

Uncertainty, the third reason for market failure highlighted by Arrow, is inherent in the research process. Companies may be unwilling to jeopardize their profitability, and sometimes their survival, by engaging in activities that have a high risk of failure. Such companies with relatively short-term horizons will, therefore, tend to underinvest in research, especially if they are unable to spread the risk over a number of projects to increase the chances of success. Again, long-range basic research will be neglected as a result of uncertainty.

Government intervention in the research system can be justified on other grounds than market failure. National defence is a government responsibility, and it would be inappropriate to allow technological developments in this vital area to be subject to the vagaries of the market and dangerous to become too dependent on foreign technology and technique. Defence research is an area in which there is much government intervention, either by procurement (government purchase) or by performance. When the government procures R & D from

private firms it is not, of course, allowing the market to dictate the rate
and direction of technical change. Government is the only 'customer'
and the nature of the end product is usually tightly specified.
Obviously, much of this procurement is applied research and experimental development but, as noted in Chapter 1, defence departments
have been, and continue to be, heavy investors in basic research too.
Meteorological services, determination of standards of measurement
and safety, public health, geological surveys and analytical services are
other areas of research activities that are appropriate government responsibilities. According to Pavitt and Walker, the economists' theory of
public goods states that when goods and services are provided for the
general public either wholly or partly independently of each individual's ability to pay for them, then there is a case for the government to
finance the R & D that is related to the provision of these goods and
services.[13]

These various arguments for public intervention in the research
system are widely accepted. The 1972 Report of the United States
President's Council of Economic Advisors said:

> Government has an appropriate role in research and development even
> when its results will not be incorporated in government purchases because
> firms would underinvest in research and development for goods normally
> purchased by the private sector. Though an investment in research and
> development may produce benefits exceeding its costs from the viewpoint
> of society as a whole, a firm considering the investment may not be able to
> translate enough of these benefits into profits on its own products to justify
> the investment. This is because knowledge which is the main product of
> research and development can usually be readily acquired by others who
> will compete away at least part of the benefits. This is particularly true of
> basic research.[14]

In its Annual Report to the Congress on Science and Technology for
the 1980 fiscal year, the United States National Science Foundation
indicated that the Federal Government's R & D investments were
focused on three areas:

- direct federal needs, where the government itself is the only or
  primary customer. Examples include national defence, space technology, air traffic control and environmental regulation;
- specific priority national needs, where the government seeks to
  augment the research efforts of the private sector because an overriding national interest demands an increase in the range of technologies available or there is a need to accelerate developments—
  energy research being an example;
- general economic and social needs, where the government assumes
  major responsibility in the national interest because incentives are
  insufficient for the private sector to invest adequately in the national

interest — federal support of basic research in medical, agricultural and educational research being typical.[15]

The Report went on to give four reasons for government intervention in research geared to the commercialization of technology. These were that technological and market uncertainties dampen private investors' enthusiasm for innovative commercial ventures; government institutions are incapable of compensating firms adequately for firms' inability to appropriate all the benefits of their investments in these risky areas; the benefits to society of federal investment exceed the costs imposed on society by that investment; and the net returns to society from the investment are at least as great as the net returns from other investments the government might make. These reasons illustrate the extent to which Arrow's three aspects of market failure have penetrated official thinking about the government's role in the research system.

But the market failure theory of intervention is not unchallenged. Whereas few would argue against government support for research that underlies the provision of adequate defence, health, education and general public welfare, intervention in the commercial sphere has its opponents. An Australian Treasury official, in discussing government subsidies for industrial R & D, is fond of quoting *Alice in Wonderland* in his opening remarks:

> When they had been running a half hour or so, ... the Dodo suddenly called out 'The race is over', and they all crowded round it, panting, and asking, 'But who has won?'. This question the Dodo could not answer without a great deal of thought, and it sat for a long time with one finger pressed upon its forehead ... At last the Dodo said, 'Everybody has won, and all must have prizes'. 'But who is to give the prizes?', quite a chorus of voices asked. 'Why, *she*, of course', said the Dodo, pointing to Alice with one finger; and the whole party at once crowded around her, calling in a confused way, 'PRIZES, PRIZES!'[16]

It does not take much effort to work out who the real-life counterparts to the fictional characters are. The runners who all got prizes are the private corporations, co-operative research associations, the universities and the professional groups of one sort or another who benefit from government intervention in the research system. Alice represents the taxpayer who pays for government handouts to the various 'contestants'. But the fact that all get prizes means there is no real contest; the competitive edge, fostered by the open market mechanism, is blunted by government largesse.

As custodians of the public purse, Treasuries, and Finance Ministries in general, are concerned to ensure that public money is not wasted and that the few are not enriched at the expense of the many. They tend to reject the arguments that the social returns from intervention far exceed the returns to private individuals and firms and that such intervention overcomes the private sector's reluctance to invest because of the uncertainty of any return.

Taking first the argument that individual firms will be unable to appropriate all the benefits of investment in R & D, the counter-argument is that the patent system (a system that is actually antagonistic to the theory of perfect competition in that it allows inventors to charge a monopoly price) is designed to ensure that a fair share of the benefits of an invention will accrue to the innovating firm.

Secondly, although there is undoubtedly a link between technological sophistication and economic growth and well-being, this does not mean that governments should subsidize firms that embark on high-cost, high-risk projects. From the producers' point of view, investment in R & D is like an investment in capital which, up to a certain point, will enable a task to be done at a lower cost than if there were less capital and more labour. Beyond that level, additional capital, whilst saving labour, will become uneconomic because its costs exceed the savings. In a similar manner, investment in R & D can reduce costs by process or product innovation. But there is a limit to the amount of research that can usefully be carried out. Eventually the point will be reached where cost reductions will be outweighed by research costs, and new products will be introduced without regard to the market's needs. The Anglo-French Concorde project provides us with one of the most convincing arguments against public support for high-risk, high-cost technological development that is subsidized for reasons other than market need. As Maxwell-Hyslop has put it, Concorde provided the airlines with an opportunity to buy an aeroplane they did not need with money that they did not have to fly routes that might not be open to them because of the nature of the technology involved. Not surprisingly, he says, the private sector airlines resisted the temptation to buy.[17] At the other end of the indivisibility spectrum, the Treasury does not favour subsidies for small producers' research needs any more than for those of the giant corporations. Small companies or even small farmers can, and do, form co-operative associations, and these should be able to provide the necessary R & D facilities and personnel. Furthermore, in the case of small manufacturers, if they have good ideas they can always sell them to larger firms which are capable of developing them to the invention and innovation stages.

Thirdly, the problem of uncertainty is, according to the critics, a fact of economic life. There is no *a priori* reason why governments should subsidize R & D just because the outcome is uncertain. The outcome of many business ventures is uncertain, oil exploration being a typical example, and this uncertainty is best handled by 'the amalgam of risk-taking decisions that we call the market.'[18] Though the market mechanism is by no means perfect, it is a better taskmaster than any government can ever be. An imperfect market will not be corrected by politicians and bureaucrats who, because they are rarely free from external influences of one sort or another, are themselves imperfect. The British and French Governments' intervention in the aircraft, computer and nuclear power industries illustrates this point very well. Governments are not in the business of 'picking the winners' out of the multitude of plausible ideas that are put forward. Winners are best picked by entrepreneurs who reap the eventual rewards if their judgement is correct; not by bureaucrats, who do not. Milton Friedman, the well-known economist and adviser to President Reagan, highlights government bureaucrats' inability to pick winners by telling the following story. When the Nobel Prizewinning nuclear physicist Leo Szilard applied to the United States Government for research grants, he always proposed to carry out experiments that he had already performed successfully. His system worked very well until one year when one of his applications was rejected on the grounds that the proposed experiment was impossible.[19]

To summarize what can be called the Treasury position, governments are motivated to subsidize industrial R & D by the fear that if they do not and other countries do, then they will be disadvantaged commercially or will appear to be backward or dependent. In the commercial sphere the fear is unfounded because, it is argued, upward movement in the competitors' exchange rate would eventually offset the advantage created by government subsidy. If governments underwrite high-cost technology of little intrinsic value for prestige or political purposes, this is both irresponsible and a waste of the taxpayers' money.

There are, therefore, two contrasting views on the need for and the desirability of government intervention in the industrial research system. On the one hand, the interventionists argue that because of the problem of externalities (indivisibility, inappropriability and uncertainty) the government must take some measures to rectify an imperfect market. These measures may include intramural performance of strategic and tactical research and direct or indirect subsidies to private firms. On the other hand, non-interventionist economists reject the market failure theory of intervention, not because the problems of indivisibility, inappropriability and uncertainty do not exist, but because they consider that the competitive free market and the patent system are sufficient to overcome them. If they are not sufficient, then it is the industrial sector, not the market, that has failed.

The linear model of innovation

None of the measures that governments have traditionally used to remedy the deficiencies of the market are free from criticism, but, as will now be evident, the measure that is most opposed is direct support of private industrial R & D. Apart from the 'Treasury position', some critics claim that the market failure theory relies on a discredited model, the linear model of innovation. This model depicts the innovation process as a linear sequence initiated either by the recognition of a need for the innovation or by a scientific discovery. The need may be stimulated by the market or there may be some other social need that the innovation would be able to satisfy; the scientific discovery may come from any of the categories listed in Table 2.1, but it is generally assumed by those who propound this model that pure research sets the process in motion. The sequence is depicted below.

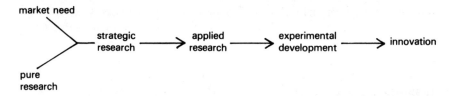

This is the sequence implied in the Bush Report and it has appeared at regular intervals in the literature since then. The British Advisory Council for Scientific Policy declared in 1967 that 'basic research provides most of the original discoveries from which all other progress flows'[20] and, in 1973, the Director of the United States National Science Foundation said:

> ... a fundamental bedrock of knowledge provided by basic research is essential to applied research oriented towards the solution of problems. Yesterday's basic research is the foundation on which today's applied research is built and, further, applied research often highlights the need for additional basic research.[21]

There is no doubt that innovations can take place in this way, for example the atomic bomb, which was stimulated by a scientific discovery, and the less dramatic Interscan microwave landing system, which illustrates very well how an innovation takes place in response to a market need. The air turbine dental handpiece is a recent example of an important technological innovation that was made without a major scientific research effort.

THE ATOMIC BOMB

It is beyond the scope of this book to describe all the scientific and technical achievements that led to the invention of the atomic bomb. It is sufficient to sketch in the briefest detail some of the important events that culminated in the discovery of nuclear fission.

Until close to the end of the nineteenth century, atoms were considered to be indestructible entities, the basic constituents of the chemical elements and unique to each element by virtue of their weights. The elements were substances that could not be broken down further by chemical action, and they entered into combination in certain proportions with other elements to form chemical compounds. Thus, sodium would combined with chlorine in equal proportions to form sodium choride (NaCl), and calcium would combine with chlorine in the proportions of one to two to form calcium chloride ($CaCl_2$).

Towards the end of the century, however, the indestructibility of the atom was questioned. Many investigators were interested in the beautiful colour effects produced when a high voltage electric charge was passed through the so-called rare gases, of which neon is one. These phenomena were called cathode rays, and J. J. Thomson at the Cavendish Laboratory in Cambridge showed that they were charged particles, much smaller than atoms. They were in fact electrons, negatively charged constituents of atoms with negligible weight. In 1895 Roentgen in Germany accidentally discovered that when cathode rays struck the walls of glass tubes, new particles, which he called X-rays, were emitted and the walls of the tubes fluoresced. It was known at that time that certain compounds fluoresced in their natural state, and the systematic investigation of many of these led Henri Becquerel to the discovery that compounds of the element uranium which fluoresced emitted radiation; they were 'radioactive'. In subsequent work Ernest Rutherford and Frederick Soddy came to the conclusion that the radiations observed when an element was naturally radioactive or when radiations were induced by bombarding the walls of glass tubes with cathode rays were of two types, alpha-particles and beta-particles. Later work showed that alpha-particles were positively charged helium atoms (of atomic weight 2) and beta-particles were simply electrons. If a radioactive element were in fact continually losing weight in the form of alpha particles, then it should be transmuting; that is, it should be decaying to some other element because the essential characteristics of an element are reflected in its weight.

The next stage in piecing together the structure of the atom occurred in 1912, when Thomson discovered two forms of neon, identical apart from a difference in weight. He called the two forms isotopes. Since chemical reactivity is associated with electric charge, there must be, he reasoned, a weight factor in the atom that had no charge but some weight. After further work with Geiger at Manchester University, Rutherford proposed a nuclear model of the atom. Most of the weight of the atom was, according to his model, located in a central, positively charged nucleus which was surrounded by moving electrons at a relatively great distance, each electron carrying a negative charge. Clearly, for neutrality, the electrical charges should balance. The brilliant Moseley then showed that nuclear charge increased in regular steps from one element to another in increasing weight order.

By 1914 the picture of the atom that had been built up was of a miniature solar system where the heavy nucleus was the sun and the light electrons were the planets orbiting at various distances from the nucleus. But there was still much to be investigated. What, for example, did the two forms of neon represent? Rutherford decided that one way to investigate the atom was to disintegrate it by bombardment with suitable missiles. He chose alpha-particles as missiles and bombarded nitrogen gas as a target. In 1917 he achieved the alchemists' dream of changing one element into another. Though their concern was to transmute base metals into gold, he transmuted nitrogen into an oxygen isotope.

In the same laboratory, F.W. Aston was conducting experiments to investigate the weights of the various elements. He found that many elements had isotopes, different forms having the same electrical charge but different weights. Importantly, Aston showed that the mass of an atom was somewhat smaller than the computed mass of the various pieces that had been identified. Since, by now, Einstein had proposed the interchangeability of mass and energy in the equation $E = Mc^2$, there was a binding energy in the atom which would be released if the atom disintegrated. Calculation showed this energy to be enormous.

Up until now, the bombardment of atoms was a rather low energy affair. The alpha-particles emitted from radioactive materials, which were used as sources, were poor atomic 'artillery'. The rate of fire was low, the sources weak and there were few ranges of energy available. Using the theoretical calculations of Gamow in America, two Cambridge physicists, J.D. Cockcroft and E.T.S. Walton, constructed a machine that would accelerate protons which could then be used as missiles. They bombarded lithium foil with these particles and succeeded in splitting the lithium atom into alpha-particles. Not only that, they calculated the energies involved and confirmed the $E = Mc^2$ equation. In the same year (1932), Chadwick at the Cavendish Laboratory discovered the neutron, the uncharged entity within the nucleus which was responsible for the existence of isotopes. Rutherford himself had speculated on the existence of the neutron in 1920, when he referred to the possible existence of an 'atom' within the nucleus of the same weight as a proton but with zero electrical charge. He pointed out that this entity, if freed from the nucleus, would be able to move through matter and enter readily into the structure of other atoms to unite with them or disintegrate them. Chadwick suggested that the neutron was an electron and a proton in close association with a binding energy of between 1 and 2 million electron volts. The atomic nucleus could now be regarded as made up of protons (positively charged: mass 1) and neutrons (electrically neutral: mass 1).

Following the achievement of a basic description of the atom, the new science of nuclear physics made enormous strides in the 1930s. Enrico Fermi, working in Italy, used neutrons instead of alpha-particles in his bombardment experiments and found that he could get excellent results if he slowed the fast-moving neutrons with paraffin wax before they struck the target. One of the target materials he used was uranium, the heaviest naturally occurring element. On bombard-ment of uranium with neutrons Fermi thought that he had obtained a new element by neutron capture. Then, in 1938, Otto Hahn and Fritz Strassman, both working in Nazi Germany, showed that the products from Fermi's experiment were barium, with a positive charge of 56 on its nucleus, and krypton, with a positive charge of 36. These products would be expected to repel each other with a force unparalleled in any atomic event so far discovered; Lise Meitner, a colleague of Hahn's, who had fled to Stockholm as a result of the Nazi persecution of Jews, calculated the force to be about 200 million electron volts. And, since further neutrons were released on fission of the uranium atom, the possibility existed for further fission until a chain reaction ensued which would release enormous quantities of energy. Early in 1939, Frederic Joliot-Curie and his team, working at the Radium Institute in Paris, demonstrated the chain reaction and published their results in the journal *Nature*. Within six years an atomic bomb was constructed, an innovation that resulted entirely from the scientific discoveries of men and women working in universities, whose motivation was the advancement of knowledge, not practical utility. It is a classic example of the linear model of innovation.

INTERSCAN

This example of the linear model is less dramatic and less far-reaching in its implications for society than nuclear energy, but it illustrates the effect that a market need can have in stimulating innovation. It con-cerns the introduction of a new type of landing system at all the major airports of the world by the early 1990s.

The aircraft landing system in use at the present time is the Instru-ment Landing System (ILS), which allows aircraft to descend to the touchdown zone along a 3° glidepath. There is only one glidepath, and it uses two broad band radio beacons which are sometimes distorted by reflections of airport buildings, parked aircraft and hills. In 1968, the International Civil Aviation Organization (ICAO) decided that a new landing system was becoming increasingly necessary because of the increase in air traffic and problems being experienced with the ILS. A series of operational requirements for the new system was drawn up by the ICAO, and member countries were asked to submit proposals for a system that would meet them. The requirements were designed to allow an aircraft to locate its position with great accuracy in three dimensions anywhere within a sector of 40° in azimuth of the approach centre runway line, up to 20° in elevation, and out to a distance of 30 nautical miles. The guidance information would have to allow the aircraft to make a curved approach of any specified form within this sector and to touch down in all weathers without the aid of a radio altimeter. To provide for missed approaches, back guidance to a distance of 5 nautical miles was also required. These specifications could not be met by any modification of the ILS; the new landing system would have to employ totally new concepts.

Proposals were submitted to the ICAO by Australia, France, the United Kingdom, the United States and West Germany. The Australian proposal was put forward by the Commonwealth Scientific and Industrial Research Organization (CSIRO). In 1971, the Division of Radiophysics began to look for areas of applied research that might complement its pure and strategic research efforts. Some urgency was lent to this quest in 1972 when the Cloud and Rain Physics Section was removed from the Division of Radiophysics, leaving the Division in the situation that its research programs were devoted almost entirely to the pursuit of pure research in radioastronomy. Being, at the time, subject to the provisions of an Act which enjoined it to carry out research for the promotion of primary and secondary industries, the Division was left in an extremely vulnerable position when the only section that could conceivably be thought of as fulfilling the Act's requirements was removed. Recognizing this, the then Chief of Division, Dr Paul Wild, convened a small committee which initiated discussions with two potential 'customers' for its work, the Department of Civil Aviation and the Overseas Telecommunications Commission.

In its discussions with the Department of Civil Aviation, the ICAO request for proposals to develop a new aircraft landing system was brought to the attention of Wild's committee. They were convinced that the proposals that had been submitted by other countries were not the best that could be achieved and that the scientists at the Division of Radiophysics could do better. By the end of 1971 a team of scientists led by Dr Wild had devised a time reference scanning beam (TRSB) system, which they called Interscan. This novel microwave system was developed in co-operation with AWA, an Australian electronics firm, and the Department of Transport. In 1974 two major contenders, Bendix Corporation and Texas Instruments, switched their allegiance from the systems they had developed (frequency encoding) to time reference scanning because of the superiority of the Australian system. Four years later, the ICAO adopted the joint United States-Australian TRSB-Interscan system.

Interscan is a typical example of an innovation that resulted from a market need and illustrates the linear model of innovation. According to Dr Wild, the basic concepts of the innovation were extracted from the literature on radioastronomy by radioastronomers at the Division of Radiophysics, which was then, and still is, one of the world's leading laboratories in this field.[22] Once the basic concepts had emerged, strategic research, applied research and experimental development followed sequentially.

THE AIR-TURBINE DENTAL HANDPIECE

Nuclear energy and the microwave TRSB landing system illustrate the linear model of innovation, and many other examples, where some or all of the steps of the sequence can be identified, could be given. But not all technological innovation takes place in a sequence that starts with basic research. Major innovations can take place with little or no prior scientific research, as the invention of the steam engine shows. In this case the theoretical principle, the second law of thermodynamics, was formulated by Carnot after the invention, not before. A more modern example of an important innovation that took place without a major scientific research effort is the air turbine dental handpiece, an invention that revolutionized dentistry when it was introduced in 1957.

The idea for a high-speed drill was conceived by John (now Sir John) Walsh, a Royal Australian Air Force medical officer, who, while testing the hearing of discharged airmen in 1945, discovered that the maximum unpleasantness of vibration sensation in teeth and bone occurred at or around the vibration frequency of the contemporary dental drill (3000 rpm). He reasoned that a drill operating at much higher speeds would create less vibration and thereby ease patient discomfort. In 1946 Walsh was appointed Dean of the Faculty of Dentistry at Otago University in New Zealand and, in collaboration with H.F. Symmons, he designed an air turbine dental handpiece with which he achieved speeds of up to 60 000 rpm. At that speed he found that enamel was removed from teeth approximately three times faster than with the conventional instrument and at one-thirtieth of the pressure.

Unable to take the development work on his bulky prototype instrument any further, Walsh turned it over to the National Physical Laboratory, a New Zealand Government research establishment. The NPL gave the project low priority and the development team was unable to overcome the problem of bearing failure which the high speeds were creating. This problem was overcome by R.J. Nelsen, a research associate at the United States National Bureau of Standards, who introduced a water-driven turbine handpiece in 1953. J.V. Borden, a colleague of Nelsen's, working first at the United States

National Bureau of Standards and then at the Dentists' Supply Company, subsequently developed the air turbine handpiece that is now known as the Borden Airotor.

Most of the credit for this major innovation should go to Sir John Walsh for recognizing the benefits that might arise from an increase in the speed of rotation of dental cutting burrs and, with Symmons, for recognizing the potential of the air turbine principle for achieving higher speeds.[23] If we are to locate the innovation within a linear sequence, it will be clear that basic research played little or no part. The main activity associated with the innovation was, first of all, the inventive activity and applied research that took place when Walsh conceived the idea of a high-speed drill and subsequently constructed the first prototype in collaboration with Symmons, and second, the experimental development that took place at the United States National Bureau of Standards and the Dentists' Supply Company which used advances in ball bearing technology to produce the successful innovation. This innovation was not stimulated by prior scientific discovery, nor can it be said that the market played a crucial role. The dental engineering companies showed no interest in Walsh's ideas and only entered the field when Nelsen conclusively demonstrated the turbine principle in the dental context. The key elements in the innovation were Walsh's recognition of an opportunity, his collaboration with an engineer in producing a prototype, transfer of the information gained and advances in a related technology.

In the three innovations discussed here, only the exploitation of nuclear energy accords with the classic linear model of innovation, but all three demonstrate the complexity of the science–technology–innovation process. In the case of nuclear energy, the key ideas arose within university laboratories from research workers engaged in the purest of pure research. These ideas were transformed into reality by a US$2 billion R & D effort. In the case of the microwave landing system, the key ideas were generated by research workers in a government research establishment, which happened to be a world leader in the pure science of radioastronomy, looking for a 'mission'. And, finally, the key idea for the air turbine dental drill came from a medical officer, untrained in research methods but with an interest in the conduction of vibration through teeth and bone, whose work paid him no dividends apart from the award of an MD degree but was, it is suggested, a key to a successful engineering innovation.

Studies of technological innovation

The complexity of the interaction between science, technology and successful innovation has not deterred research teams from undertaking numerous case studies to try to unravel it. It is beyond the scope of this book to review the entire research effort, but some of the studies that bear upon government intervention in the research system will be described.

In one of the earliest studies of the sources of invention, Jewkes, Sawers and Stillerman conducted an historical survey of 61 major twentieth-century inventions.[24] These included Bakelite, catalytic cracking of petroleum, cellophane, the ballpoint pen, magnetic recording, streptomycin, insulin, the jet engine, the helicopter, radio, xerography and the electron microscope. Many, like Terylene, polythene and DDT, were household names. These authors concluded that the sources of invention were numerous, scattered and varied. Most of the inventions were due not to the research laboratories of the big corporations, but to individual independent inventors working either in a university or on their own account. The pattern of invention had, therefore, not significantly changed in the twentieth century. Despite the rise of the industrial R & D laboratory, it appeared that invention was still something that could not be planned and was as prone to the 'slings and arrows of outrageous fortune' as ever. Out of the 61 inventions, 33 were due to individuals, 21 occurred in industrial research laboratories or were due to mixed government–industrial research activity, and the remaining 7 were not classifiable. Jewkes and his colleagues concluded that

> the theory that technical innovation arises directly out of and only out of advance in pure science does not provide a full and faithful story of modern invention. As in the past three centuries there is still a to-and-fro stimulus between the two.[25]

In a later edition of their work, which added seven more case studies to their original list, Jewkes and his colleagues reiterated this view. 'There are grounds for doubting,' they said, 'that science and technology always march together; whether science, that is to say, always carries with it an economic pay-off.'[26]

The Jewkes, Sawers and Stillerman study has been described by Professor Christopher Freeman, Director of the Science Policy Research Unit at Sussex University, as a classic in science policy research which has had enormous influence in the United States and Western Europe.[27] He criticizes it on a number of counts, however. First, he does not agree that the 61 major innovations the team selected are representative of successful twentieth-century inventions. A vast number of smaller, less spectacular inventions that improve upon existing products and processes were ignored. These, too, are necessary to achieve technical progress and economic growth and, taken together, are as important, if not more so, than the spectacular breakthroughs. Second, the list is biased in favour of the independent inventor in a number of ways. For example, more than half of the inventions selected were made before 1928, when there were relatively few industrial laboratories. Twenty of the individual inventions took place before 1928, whereas only 8 took place after that year. On the other hand, 8 corporate inventions took place before 1928 and 19 afterwards. In the area of synthetic materials, plastics, which were produced in great quantity and resulted from corporate research, were ignored in favour of a list that contains 2 individual and 5 corporate inventions. And finally, bias is evident in one specific case study, the catalytic cracking of the heavy fractions of crude oil, in which major credit is given to an independent inventor despite the fact that his process never accounted for more than 5 per cent of refinery and world output and was soon superseded by a vastly superior process developed by the big corporations.

In a later edition of their book, Jewkes and his colleagues were unrepentant. Indeed, they extended the policy implications of their work by suggesting that 'it can no longer be claimed that supremacy in science guarantees maximum economic growth'. This is a theme that had been developed by Professor B.R. Williams a decade before, when he purported to show, by international comparison, that the countries that spent the most money on R & D were not thereby guaranteed a high rate of economic growth.[28] The correlation that Williams obtained between expenditure on R & D as a proportion of Gross National Product (GNP) and per capita GNP in the same year (1961) is shown in Figure 2.1.

It will be seen from the figure that, in 1961, Australia, France, West Germany, Sweden and the United Kingdom had about the same per capita GNP, but their R & D expenditures as a proportion of GNP ranged from 0.6 per cent to 2.5 per cent. Williams then plotted the annual growth rates in per capita GNP in the years 1951-60 against the R & D expenditures of the various countries over the same period. This correlation is shown in Figure 2.2.

Williams' results show that there is no real connection between expenditure on R & D and national wealth as measured by the GNP. It might be argued that since the richest country (the United States) spends the greatest percentage of its GNP on R & D and the poorest (India) spends the least, there is a causal relationship between the two. But this does not explain why the United Kingdom, with the second highest expenditure on R & D, had about the same per capita GNP as Australia, which spent a quarter of that amount as a proportion of its GNP. All that can be concluded is that the rich countries can afford to spend more on R & D than poor ones, and that there is not necessarily a causal connection between a nation's wealth and its expenditure on R & D. The results tell us that Japan, with the highest annual economic growth rate in the period 1951-60, spent relatively little on R & D, whereas the United States, with one of the lowest annual economic

Figure 2.1

GNERD vs per capita GNP

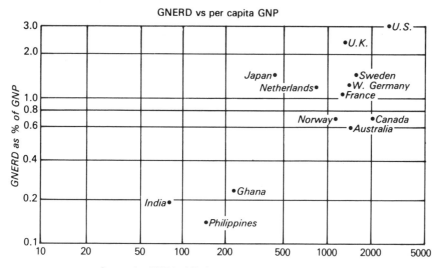

Per capita GNP in US dollars at year of R.D. estimate

Source: B.R. Williams, 'Research and Economic Growth — What Should We Expect?', *Minerva*, 111(1964)

Figure 2.2

R & D vs rate of growth

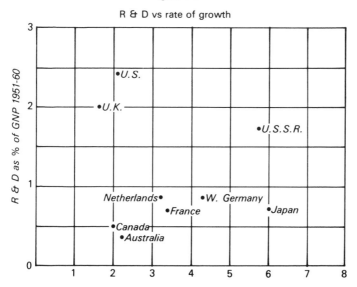

Annual % growth of national product per man-year, 1951-60

Source: B.R. Williams, 'Research and Economic Growth — What Should we Expect?', *Minerva*, 111(1968)

growth rates of the group, spent the highest proportion of GNP on
R & D.

The Williams data do not tell us the amount of the gross expenditure
on R & D that was devoted to defence and space activities. This
expenditure relates to the GNP in quite a different way from that
directed to civil commercial ends. Japan, for example, undertook very
little defence research, whereas an overwhelming proportion of the
United States and United Kingdom research expenditure was in this
area. It was assumed in the study that expenditure statistics were
directly comparable and that the expenditure would be reflected in
economic growth within the ten year period. Despite these limitations,
however, the Williams study did show that economic growth is not
likely to be increased simply by increasing expenditure on R & D.

After publication of the Jewkes study, investigations into the role of
R & D in economic growth gathered pace. In 1961 D.J. de Solla Price
suggested, on the basis of historical evidence, that only rarely does
science give rise to technology directly and that the time lag between
discovery of knowledge and its application was not getting shorter,
conclusions that ran counter to current thinking. 'It is naive,' he said,
'to regard technology as applied science. Because of this one should
beware of any claims that particular scientific research is needed for
particular technology potentials.'[29]

In 1965, the United States Department of Defense (DOD), believing
that the time had come to make a comprehensive analysis of the
impact of research and technology on a substantial number of weapons
systems in use, in procurement or at an advanced stage of engineering
development, established Project Hindsight to

- identify and firmly establish management factors for research and
  technology programs which have been associated with the utilization
  of the results produced by these programs; and

- measure the overall increase in cost effectiveness in the current
  generation of weapons systems compared to their predecessors which
  is assignable to any part of the total DOD investment in research and
  technology ($10 billion since 1946).

The project team, led by Colonel Raymond Isenson, adopted a four-
fold strategy in carrying out the investigation. They intended to

- determine the extent to which new weapons systems depended on
  the results of recent advances in science or technology for the
  increase in system effectiveness, decrease in costs, or increase in cost
  effectiveness as compared to the previous system which they
  attained;

- determine the proportion of any new technology required for attaining characteristics of the system that resulted from DOD-financed research in science and technology;

• determine significant management and other environmental factors, as seen by the research scientist or engineer, that appear to have contributed to high utilization of research results;

• quantify the return on DOD investment in research.

The basic element of quantification the team adopted was the Research or Exploratory Development Event (RXD), which they defined as a scientific or engineering activity during a relatively brief period of time that included the conception of an idea and the initial demonstration of its feasibility. In an innovation there could be one or a series of RXD events, culminating in a device or component found in the weapon system under study.

In its final report, issued in 1969, the Hindsight team revealed that, of the 686 RXD events that were identified in their study of 20 weapons systems, only 9 per cent were due to research. The precise breakdown is shown in Table 2.2. The team concluded that there was very little evidence in their study to support the view that ideas generated during recent basic research, and especially pure research, were carried forward into the 'applied research and incentive activity' state of innovation or to final developmental work. Only a few examples were found — in statistical analysis, nuclear physics and polymer chemistry.

Table 2.2
Research events of Project Hindsight

Activity	Events as % of total
Pure research	0.3
Applied research	8.7
Exploratory development	28
Advanced development	40
Engineering development	22
Management and support	1

When the interim report of the Hindsight team was issued in 1966, the academic community regarded its conclusions about the role of basic research as totally disastrous. With DOD support for basic research at the time running at about US$100 million a year, it was disconcerting to read in the report that

> Science and technology funds deliberately invested and managed for defense purposes have been about one order of magnitude more efficient in producing useful events than the same amount of funds invested without specific concern for defense needs.[30]

Shortly after the publication of the interim report, the National Science Foundation invited proposals for a 'study to investigate the manner in which non-mission-related research has contributed over a number of years to practical innovations of economic or social importance'. The resulting study, entitled 'Technology in Retrospect and Critical Events in Science' (TRACES), was carried out by a team from the Illinois Institute of Technology using a critical events methodology and an historical cut-off point of fifty years.[31] Five major twentieth-century innovations were chosen for study — the videotape recorder, the oral contraceptive pill, the electron microscope, magnetic ferrites and matrix isolation, with the following results:

- in all the case studies, pure research provided the origins from which science and technology could advance toward the innovations that took place;

- of the key events documented, approximately 70 per cent were classified as non-mission (pure) research, 20 per cent were mission-oriented (strategic and applied) and 7 per cent were development and application;

- the distribution by performers of key events was as shown in Table 2.3.

Table 2.3
Research events of TRACES

	University and college	Research institutes and government laboratories	Industry
Non-mission research	76	14	10
Mission-oriented	31	15	54
Development and application	7	10	83

Source: IIT Research Institute, *Technology in Retrospect and Critical Events in Science* (Washington, 1968)

These results appeared to re-establish the supremacy of the linear model of innovation and to overthrow the Hindsight conclusions about the role of pure research in innovation. Subsequent investigations have attempted to add to our knowledge of the relationships between science and technology and economic growth by adopting different methodological approaches and more random sampling techniques. Two British studies are of note — Project SAPPHO, carried out by the Science Policy Research Unit at the University of Sussex, and the Manchester University Queen's Award Study.

Project SAPPHO was an attempt to systematically identify and evaluate the factors that distinguish innovations that have achieved commercial success from those that did not. Drawing its examples from the chemicals and scientific instruments industries, 29 pairs of innovations were studied, 17 in chemicals and 12 in scientific instruments. Each pair consisted of one commercially successful innovation and one failure, both aimed at the same market. Innovations were regarded as failures if they did not obtain a worthwhile market share or did not make a profit, even if they worked in the technical sense. In the case of successful innovations,

- there was a much better understanding of user needs;
- much more attention was paid to marketing;
- the development work was performed more efficiently than in the case of failures, but not necessarily more quickly;
- more effective use was made of outside technology and scientific advice, even though most of the work was done in-house;
- more senior personnel were in charge of the projects.[32]

The SAPPHO team confirmed the view that organized R & D was characteristic of innovation in the two industries studied. Indeed, competent R & D was regarded as a *sine qua non* of innovation in the chemical industry. Most of the innovations were carried out within the firms' research departments, and though the existence of a research department or facility was not a guarantee of success, it was a prerequisite for attempting to innovate. The team failed to find substantial evidence to support the view that successfully innovating firms conducted a certain amount of basic research in their laboratories.

The second major British study of technological innovation was published in 1972 by a team of researchers at the University of Manchester, using the Queen's Award Scheme to select case studies.[33] The Queen's Award to Industry is granted to individual companies for export achievement or technological innovation or both, and, in the years 1966 and 1967, 66 awards were made involving 84 technological innovations. These innovations formed the basis of the study, making it the most wide-ranging innovation study carried out to date.

The Queen's Award team divided the linear model of innovation into two types which they called the 'discovery–push' and the 'need–pull' models. In the discovery–push model, a scientific or technological discovery set the innovative process in motion; in the need–pull model, a customer or a management need provided the stimulus. The team found that very few of the innovations studied fitted either of the two models perfectly. Significantly, none of the 84 innovations occurred as the result of a scientific discovery, and only 5 resulted from a technological discovery. It was not possible, therefore, to identify any basic research discoveries whose timing was crucial for any of the innovations. Furthermore, in tracing the sources of the 'major technical ideas' that resulted in the innovations, only 10 out of a total of 158 came from the universities.

The team concluded that the transition from pure knowledge to economic payoff is less simple, and less direct, than was commonly supposed at the time. The benefits that they saw arising out of pure research were that it provided techniques of investigation, the people trained in using those techniques, and innovation already embodied in the form of new instruments developed during the course of pure research. But on the matter of straightforward economic benefit from the tangible output of pure research, their opinion was uncompromising. 'It is difficult to believe,' they said, 'that much national economic benefit arises from the knowledge output of a nation's own basic research.' The academic community could derive little comfort from this bleak assessment and still less from the subsequent work of Langrish, the leader of the team, who singled out university research for further examination.

On the assumption that industrial scientists who are asked to review developments in their fields will cite the university research which they consider to be of relevance, Langrish analysed seven review articles written by British industrial chemists in the 1967 volume of *Reports on the Progress of Applied Chemistry*. The reviews covered developments in the following fields: phenolic resins, microbiological techniques in manufacture, water treatment, plastics, polyolefins, synthetic detergents and poultry and eggs. The results of the analysis showed that most of the papers which the reviewers considered most relevant to technological developments in these fields did not come from universities. Eighteen per cent came from universities, 34 per cent from industry, 16 per cent from government sources and 30 per cent were untraceable (see Table 2.4).

From these data, Langrish concluded that the small contribution of ideas from university science to innovations and the lack of interest shown by industrial chemists in the publications of academic chemists would make it more difficult for policy-makers to think that basic research provided most of the original discoveries from which all other progress flowed.[34] The United Kingdom Council for Scientific Policy had said precisely this about the origins of progress in its *First Report on Science Policy*, issued in 1967.

Langrish's conclusion about the small contribution of ideas from university science to innovations can be challenged, however. The areas chosen for analysis have not been shown to be representative of the areas in which university scientists work, and it is likely that they are not. It is hardly surprising to find that citations to research activity in 'plastics' or 'detergents' from university research laboratories are rather few. Without some indication of the amount of university work being carried out in each of the fields analysed, the data are

Table 2.4

Institutional sources of cited papers in seven reviews
in *Reports on the progress of applied chemistry*

Review	Industry		Government		University		University industry collaboration		Not available
	U.K.	Abroad	U.K.	Abroad	U.K.	Abroad	U.K.	Abroad	
Phenolic resins	3	19	2	4	0	0	0	1	29
Microbiological techniques	6	2	2	2	7	2	2	0	14
Water treatment	9	38	27	34	5	45	2	0	42
Plastics	5	1	0	0	0	0	0	0	4
Polyolefins	12	71	0	0	2	12	3	2	41
Synthetic detergents	1	27	4	1	0	1	0	0	17
Poultry & eggs	2	0	10	6	2	19	0	3	24
Totals	38	158	45	47	16	79	7	6	171
Totals as % of 567	7%	27%	8%	8%	3%	13%	1%	1%	30%
Total as % of 396	10%	40%	11%	12%	4%	19%	2%	2%	—

Source: J. Langrish, 'Technology Transfer: Some British Data', *Research and Development Management*, 1(1971), p. 135

Table 2.5

Institutional sources of cited papers in five biological reviews in *Reports on the progress of applied chemistry*

Review	Industry		Government		University		University industry collaboration		Not available
	U.K.	Abroad	U.K.	Abroad	U.K.	Abroad	U.K.	Abroad	
Prostaglandins	0	5	0	4	7	34	0	5	4
Enzymes	2	34	6	26	9	51	1	5	18
Microbial insecticides	0	2	2	20	2	11	0	0	3
Antiviral agents	5	15	1	6	9	10	0	1	3
Anabolic steroids	0	12	0	1	0	4	0	2	6
Totals	7	68	9	57	27	110	1	13	34
Totals as % of 326	2%	21%	3%	18%	8%	34%	0%	4%	10%
Totals as % of 292	2%	23%	3%	19%	9%	38%	0%	5%	—

Source: G. Pragier and J. Ronayne, 'A Criticism of the use of Citation Analysis in Studying the Science-Technology Relationship', *J. Chem. Information and Computer Sciences*, XV(1965). p. 156

meaningless. In the one area where citations to university research dominate — microbiological techniques — the result indicated to Langrish that the main use of university research is to provide techniques for industrial research workers, an explanation that is in line with the conclusions drawn in the Queen's Award Study. An alternative explanation is that the nature of the subject will dictate the usefulness or otherwise of university research to industry. For example, a study of citation patterns in biologically oriented reviews in the same journal that Langrish used shows that references to university work outstrip those from other sources.[35] The results are shown in Table 2.5. Different results could be expected in reviews related to physical, chemical and engineering sciences.

As we have seen, Langrish emphasized the policy implications of his citation analysis. While one cannot be certain that policy-makers have been influenced by the results of the various studies of the science–technology relationship, it certainly appears as if some have. For example, the United Kingdom Council for Scientific Policy, in its third report, did a remarkable about-turn and declared: 'Curiosity-oriented (i.e. pure) research is only rarely the mainspring of substantial innovation.'[36]

The market failure theory and government intervention

The most important conclusion to be drawn fom this review of the sources of invention and innovation is, echoing Jewkes, Sawers and Stillerman, that the sources of invention are numerous, scattered and varied. The linear model of innovation should, therefore, not be taken as the only description of the inventive and innovative processes. Innovation can take place in a linear sequence, starting out as pure research in a university or college laboratory and finishing up as a new or improved product or process, but this is the exception rather than the rule. This is not to say that there are no serious doubts about the selection procedures and methodologies of some of the studies that purported to overthrow (or support) the linear model. Project Hindsight has been described by Kreilkamp, with good reason, as inept and biased,[37] and Layton, in reviewing the findings of both Project Hindsight and TRACES, invites us to suspend scholarly judgement because 'there is more than a suspicion that the interests of sponsoring agencies influenced the results.'[38] Even the Queen's Award Study is not immune from a suspicion of bias. By defining research in institutional terms, that is, 'university research', 'industrial research', 'government research' and so on, and then choosing case studies on the basis of an award designed to recognize innovations in areas that one would not expect to be heavily represented in university research laboratories (shuttle-less looms, aggregate from pulverized fuel ash, industrialized building, for example), the team may have loaded the dice against the universities.

The case studies used have not been systematically evaluated, but in one case, the discovery of the semi-synthetic penicillins by the Beecham Company, an evaluation by Pragier revealed a tendency on the part of the Queen's Award team to downgrade the role of university research. The isolation of 6-aminopenicillanic acid (6-APA) from fermented liquors and the subsequent conversion of the acid by the Beecham scientists to wide spectrum acid-resistant and penicillinase-resistant 'semi-synthetic' penicillins was one of the most significant advances made in the production of chemotherapeutic agents in this century. In considering this innovation, the Queen's Award team said:

> The preparation of new penicillins was made possible by a discovery. This discovery, the isolation and preparation of 6-APA, was scientific, but it was not the product of pure science . . . Most of the 'academic science' made use of by Beecham scientists in their work was of distant origin.[39]

Natural penicillin was discovered by Alexander Fleming, working at St Mary's Hospital Medical School in London in 1928. In subsequent work at the University of Oxford, Chain, Florey and Abraham developed the crude antibiotic to the stage, in 1941, where human trials could be carried out. The work leading to the trials was the only contribution from 'academic science' that the Queen's Award team recognized, and even that recognition was tempered by the inaccurate statement that the work was motivated by a desire to improve the health of the fighting forces during the Second World War. Sir Ernest Chain had this to say about a year before the Queen's Award team published their work:

> I should like to point out that the possibility that penicillin could have practical use in clinical medicine did not enter our minds when we started our work on penicillin. A substance of the degree of instability which penicillin seemed to possess according to the published facts does not hold out much promise for practical application. I started work on penicillin in 1938, long before the outbreak of the war. The frequently repeated statement that the work was started as a contribution to the war effort, to find a cheap chemotherapeutic agent suitable for the treatment of infected war wounds, has no basis. The only reason which motivated me to start the work on penicillin was scientific interest. I very much doubt, in fact, whether I would have been allowed to study this problem at that time in one of the so-called 'mission-oriented' practically-minded industrial laboratories. The research on penicillin, which we started as a problem of purely scientific interest but had consequences of very great practical importance, is a good example of how difficult it is to demarcate sharp limits between pure and applied research.[40]

In 1954, Chain, acting as a consultant to Beecham's, advised the company to attempt to make penicillin by semi-synthetic means, and it was as a result of this advice that the research team which discovered 6-APA was established. This stimulus, from an academic researcher, does not, and indeed cannot, according to the methodology, appear as a key idea originating in the universities. Subsequently, the Beecham claim to the discovery of 6-APA was challenged by the Bristol Laboratories, a subsidiary of the United States-based Bristol-Myers Corporation, on the grounds that work carried out in Japanese universities by Sakaguchi and Murao had foreshadowed the Beecham isolation of 6-APA to such an extent that the company's work 'did not provide an exercise for any inventive ingenuity.'[41] Beecham, in reply, asserted that although the Japanese workers had published work which contained the structural formula and some properties of 6-APA some two years before their own breakthrough, these properties did not coincide with those of the Beecham material. Judgment was given against the Beecham Group in the British High Court in 1967, but this decision was reversed on appeal to the Supreme Court.

The litigation demonstrates that the major inventive step in the semi-synthetic penicillin innovation was not so clear-cut as the cursory examination contained in the Queen's Award Study suggests. Bristol Laboratories were willing to test it out in court and only lost when the case went to the Supreme Court. Pragier, using this and other evidence, suggests that the Beecham discovery was made against a background of previous research, largely academic, some of which was overlooked by the Beecham researchers. To interpret the story as one in which the key ideas all came from industry is, according to him, a distortion brought about first of all by ignoring the academic work that went on in the 1950s in both the United States and Japan; secondly, by adopting an arbitrary cut-off point which dismissed the work of Fleming, Chain and Florey; and thirdly, by the fact that the methodology is unable to accommodate the influence of an academic consultant whose advice set the innovating firm off on its path to profit.

Whether there is evidence of such bias against the universities in the other Queen's Award case studies is not known. In subsequent work at the University of Manchester, the universities fared a little better when Gibbons and Johnston examined the information inputs that contributed to solving technical problems in thirty British innovations involving new products. None of the innovations, which were drawn from the new products pages of trade journals, were generated by a scientific discovery, but of the 887 inputs that the authors identified, 107 were classified as scientific; that is, they were due to either personal contact with or reference to the literature of people recognized as scientists.[42] Academic scientists and their literature accounted for at least 60 of these scientific inputs. While this might not seem very much, it was at least a better result for university research than the Queen's Award Study. Gibbons and Johnston concluded that:

> ... it is apparent that the relationship between science and industrial technology is more complex than previously assumed by either scientists or economists; there exists a wide variety of potential forms of interaction. While this settles the issue of whether science contributes to technological innovation, and provides a justification for maintaining an effective research capability, the very complexity of the relationship precludes simple calculations of the optimum size or distribution of the science budget.[43]

Like the other studies of innovation, however, it is possible that bias was introduced at the sampling stage and it is always possible that entirely different results might be obtained from a different sample, even from the same journals. As Freeman has pointed out, 'generalizations arising out of innovation studies need to be heavily qualified. The universe of invention and innovation is not known and no strictly random sample can be drawn.'[44] The Gibbons and Johnston study was important, however, in that it confirmed the Jewkes view that the sources of invention and innovation are numerous, varied and scattered. Technical change is a complex process involving interactions and interdependencies across the whole spectrum of scientific and industrial activity and is strongly influenced by economic and other government policies.

The market failure theory and the linear model

In a review of the techniques governments use to promote industrial innovation, K.G. Gannicott rejects the market failure theory and the policy instruments used to overcome it because of their supposed reliance on a linear model of innovation.[45] It has previously been shown that the government schemes aimed at improving the competitiveness of science-based industry are justified in terms of the market failure theory and usually involve some sort of R & D support. It is assumed that if the level of R & D is increased, there will be a corresponding increase in technological innovation, and government industrial policy must, therefore, include measures aimed at achieving an appropriate balance between basic and applied research. Gannicott argues that the assumption, and the policy that flows from it, reflects an unquestioning belief in the unsatisfactory linear model of innovation. The typical process of technical change is not, he says, the dramatic breakthrough in the laboratory but incremental improvement to existing products and processes brought about by engineers, production supervisors and designers without any direct contribution from scientific research. The market does not fail. It stimulates the search for these incremental improvements and is responsible in any case for the majority of industrial innovations. On these grounds there is a strong case for abandoning the market failure theory and the policies towards industrial innovation that flow from it. The maintenance of large national laboratories devoted to the pursuit of basic research and direct support for industrial R & D is, according to this view, a waste of public money, especially in small countries which really cannot afford the luxury of supporting a relatively useless research effort.

The linear model of innovation is, of course, misleading if taken as the only interpretation of the innovative process. A great deal of technological innovation does take place as a result of market demand, and governments should take note of this when they formulate policies to stimulate industrial innovation. The technological sophistication and innovativeness of American industries engaged in activities relevant to defence and space are due in no small measure to extensive government procurement and contracting. Gannicott suggests that the customer–contractor principle, introduced by the United Kingdom Government in 1972, is a technique whereby the R & D system might be stimulated to respond to the needs of administrative departments; the 'market', represented by the departments, commissions appropriate R & D from the Research Councils. This customer–contractor principle will be discussed again in Chapter 5. In considering their policy instruments for stimulating industrial innovation, governments committed to the market failure theory should formulate remedies that are appropriate to the type of failure that they are trying to counteract. As Gannicott rightly points out, subsidies are not the answer to all the elements of the market failure theory and alternatives are available which should be tried. For example, loans, or venture capital, granted under specific conditions might well be a more effective method of remedying uncertainty than direct subsidy. Subsidy is, theoretically, the appropriate mechanism for overcoming inappropriability, provided that significant benefits to the community are expected.

The suggestions that the market failure theory must be abandoned as a justification for government intervention in the R & D system because the linear model provides an inadequate description of the innovative process and because a great deal of industrial innovation apparently takes place as a result of market forces must be considered carefully.

Let us take up the matter of the linear model of innovation first. The sources of invention and innovation are scattered, numerous and varied and any description of the innovative process which assumes that this process always starts with basic research is simply wrong. Some innovations clearly do have such a starting point, however, and in allowing basic research to be neglected, as it would be if left entirely to the private sector according to the market failure theory, governments would be taking a conscious decision to forgo the sorts of public benefits that have in the past arisen from basic research.

Though the proportion of basic research that leads to innovation may be quite small, many believe that governments simply cannot afford to neglect it because, as even Project Hindsight observed, '. . . it has been seen that wherever there was a pay-off of basic scientific research in terms of use in systems, its value was comparatively great.'[46] Even if, for the sake of argument, the possibility of direct economic pay-off from basic research is ignored and, further, it is assumed that applied research and experimental development in industrial and government laboratories are responsible for most of the technological innovation that results in economic pay-off, who has trained the research workers who contribute to this pay-off, and how was the training carried out? Most of these research workers have, of course, been trained in the universities and other institutions of higher learning, and generally their training involves the pursuit of basic research, either pure or strategic. The work of Schmookler has been quoted in an attempt to downgrade the role of university-trained research workers in technological innovation, but this is not convincing. Schmookler identified the educational background of a sample of United States patentees who had filed patents over a two month period in 1953 and found that a low proportion of the inventors had tertiary qualifications.[47] While it is impossible to generalize from such a limited study, Jewkes, Sawers and Stillerman's extensive study showed that many of the individual inventors to whose activities these authors gave great prominence were either working in universities or were university trained. It can be concluded, therefore, that although the linear model is not the only description of the innovative process, it does not necessarily follow that the market failure theory cannot be used to justify government support of research.

The argument that government intervention in the research system is unwarranted because a great deal of technological innovation is stimulated by market demand and is generally the result of incremental improvements by production personnel is also unsound. If the evidence on which such statements are based were incontrovertible, then there is no doubt that the wisdom of government intervention in research would be seriously in question. But the evidence is not incontrovertible. It is true that Schmookler and others showed that, in the industries they studied, minor improvements and new designs were as important for technical progress as the breakthrough innovations.[48] But there was no suggestion in their work that these incremental improvements were largely the result of innovations generated by production personnel rather than the professional R & D system. On the contrary, in his study of innovations in Du Pont's rayon plants, Hollander found that the major technical changes were largely developed by the company's formal research groups and that the minor technical changes, which he considered to be as important as the major ones, were due to Technical Assistance Groups—trained chemists and engineers concerned with solving problems.[49] It cannot be readily accepted, therefore, that technical change is largely the result of incremental improvements by production personnel. The weight of evidence to the contrary is overwhelming.

If it were true that the initial stimulus for the majority of technological innovations comes from market demand, then one would expect that the necessary R & D would be carried out without any help from government. Again, it cannot be said that the evidence for such a suggestion is incontrovertible. In one of the most thorough critiques of the various studies that claim to demonstrate the primacy of market demand forces, Mowery and Rosenberg concluded that the notion that market demand forces 'govern' the innovation process is simply not demonstrated by the empirical analyses which have claimed to support that conclusion. 'The studies,' they say, 'may be criticized at the specific level of failing to substantiate their hypotheses; the primacy of market demand forces within the innovation process is simply not demonstrated.'[50] Project Hindsight, TRACES, the Queen's Award Study, Project SAPPHO and the work of Marquis and Myers[51] and Gibbons and Johnston were included in the Mowery and Rosenberg critique. Some of the more serious criticisms were:

- The studies, in general, focused on 'producer goods' rather than 'consumer goods' innovations. In supplying goods to other producers, manufacturers have to work towards particular characteristics and specifications demanded by the customer. Attention to users' needs would be expected, therefore, to be of greater importance than in innovation aimed at the less sophisticated customers in consumer goods. In that sense, one would expect that the 'market' would have an important role to play in initiating the necessary R & D.

- Only successful innovations were studied (except in the SAPPHO project) and by definition there must be a demand for the product or process. In the case of Marquis and Myers, who studied 597 successful innovations and concluded that 45 per cent were due to market factors, Mowery and Rosenberg critically remark, 'All that (they) established is that there was an adequate demand for those innovations which turned out to be successful. We agree, but how would we disagree?'. In Marquis and Myers' scheme, demand could be either current or potential. Since potential demand exists for almost anything, it becomes virtually meaningless.

- In general, the definitions of 'needs' and 'demands' were very loose.

In some of the studies, the authors appeared to be unaware of the distinction between supply and demand. Thus, if a firm commits resources to R & D because of pressure from competitors, the stimulus is being provided from the supply side, not from market demand. Similarly, Langrish and his colleagues defined as a 'need–pull' a manufacturer's need to reduce the costs of a manufacturing process or the need to avoid a take-over, either of which may induce the manufacturer to commit resources to R & D. This need–pull has little to do with market demand; it is a supply side phenomenon, yet in interpretations of the Queen's Award Study, even by Langrish himself, instances such as these are attributed to 'demand–pull'.

- The importance of the various innovations studied was never assessed systematically. There are occasional hints that when scientific or technological discoveries give rise to an innovation, this is a more important innovation than those that are stimulated by what the authors define as demand. There was such a suggestion in Project Hindsight; and in the Queen's Award Study, too, it was claimed that for major innovations the recognition of a discovery's usefulness was found to be more important than was true for the total sample. But this phenomenon is not systematically followed up.

- There is, in general, a failure to distinguish the factors that are important in stimulating an innovation from those that are important in diffusing it. Mowery and Rosenberg suspect that, if diffusion factors could be taken into account then the results might be different. In diffusing an innovation — that is, making it a success in the market — the influence of market demand would naturally be paramount.

- Twenty-three per cent of the 567 innovations Marquis and Myers studied were existing innovations acquired by a particular firm. In studying the firm's subsequent development of the innovation, the underlying science and technology that led to the innovation will not show up, and results are biased in favour of market demand.

On the basis of these and many other criticisms, Mowery and Rosenberg concluded that:

> The role of demand has been over-extended and misrepresented with serious possible consequences for our understanding of the innovative process and of appropriate government policy alternatives to foster innovation. Both the underlying evolving knowledge base of science and technology, as well as the structure of the market demand play central roles in innovation in an interactive fashion, and neglect of either is bound to lead to faulty conclusions and policies.[52]

In view of these powerful criticisms, it would be foolish to abandon the market failure theory as a guideline and a justification for government intervention in the R & D system. Market forces will generally ensure the rapid diffusion of worthwhile innovations with good profit potential, but it has not yet been demonstrated that such forces can be relied upon to initiate the steps leading to the innovations. Langrish and his co-workers have acknowledged that technological innovation is a synthesis of some kind of need with some kind of technical capability. Need alone cannot guarantee the existence of technical capability; nor, according to the market failure theory, can the private sector. The government properly intervenes at the research end of the innovation spectrum, where it can support and complement industry and foster social benefits which would not result from the market system alone. It is not the only way that governments can influence innovation, but it is an important one.

CHAPTER 3

The uneasy alliance

Few governments of developed or developing countries do not now intervene in the funding and performance of scientific research dedicated to social, economic or political objectives or the advancement of knowledge for its own sake. As there are no convincing arguments to the contrary, the market failure theory can justify government intervention and offer some guidelines as to how financial resources should be distributed. However, the market failure theory does not indicate what level of government support is appropriate, how resources should be allocated between the various sectors, or what criteria should be applied in making decisions about resource allocation.

In 1971, the United States National Institutes of Health commissioned Professor R.A. Rettig to review the literature of the previous twenty years dealing with criteria for allocating financial resources to R & D in the public sector. He was to analyse published criteria for determining the appropriate level of research expenditure, particularly on biomedical research; for allocating resources along the continuum of fundamental (i.e. basic) research, applied research, development, testing and demonstration; and for allocating research funds among domestic program areas and hence for allocating research funds among disease categories.[1] Though the National Institutes of Health were mainly interested in biomedical research, their reaction to Professor Rettig's report is relevant to all areas of scientific research:

> An outstanding feature of this growing body of literature [on criteria for choice] is the absence of any intellectually convincing or satisfying algorithm for arriving at an acceptable or optimal level of expenditure for current or projected budgets. The determination of such a level in the event that science were to be supported for non-utilitarian purposes — for its contribution to the cultural level of the Nation or for the gratification of the needs of a small segment of the society — would be difficult indeed. But even when the context is utilitarian, a transformation of a social objective into dollar levels of support for research has remained intractable.[2]

More than a decade later, the problem is still with us. According to traditional economic theory, public resources should be devoted to R & D until such time as the marginal social utility arising from other forms of investment exceeds that arising from investment in R & D. But because of the uncertainty of pay-off from research, the information needed to perform this calculation is not available. Indeed, it is very difficult to calculate the rate of return on investment in research that has led to the production of goods and services. When this is attempted, the results are at best unreliable and at worst grossly misleading. In determining how much support should be given to the various categories of R & D, too, economics offers little guidance. It is useless to know that the criterion is that public money should be invested up to the point where the private sector becomes willing to invest in research, since the private sector's willingness to invest depends on a number of factors. For example, as the degree of uncertainty and inappropriability varies, so too does an individual firm's perception of the potential return on investment. The potential size of the market is often a crucial consideration, and different firms may perceive this differently. Government regulations in a particular area may act as a disincentive to invest in research, a claim which is made by the pharmaceutical industry. It is therefore almost impossible to foretell what decision a firm or an industry will make with regard to investing in research.

Given that the level of government support needed to maximize the social benefits of investment in R & D is not known, what course of action is open to government decision-makers? There has been a general tendency to evade the issue by adopting rule-of-thumb procedures based on those of the very recent past, with adjustments to take into account the current economic situation or overall government economic policies. Such rules-of-thumb may be applied to target percentages of the GNP that should be spent on R & D in the nation as a whole and to the percentages that should be devoted to the various categories of research. These percentages are often based on international comparisons, the big spenders usually being used as the benchmark.

In the 1960s and 1970s even the big-spending nations reduced the proportion of GNP that they devoted to R & D, and this made scientists and administrators aware of the need to establish criteria for evaluating scientists' claims on the public purse against those of other sectors of the economy and also to identify priority areas within R & D.

According to Harvey Brooks, however, the pressure to develop such criteria arose for other than financial reasons. Firstly, there is the belief that, since science is an inherently 'rational' activity, scientists should be capable of formulating rational, objective criteria by which choices might be made between competing fields and within the fields them-

selves. There are also areas of scientific activity that must be judged in
their own right and not by reference to external considerations. The
criteria for making such judgments had not been considered before the
so-called scientific choice debate of the early 1960s. Secondly, the
priority given to many basic disciplines after the Second World War
resulted from their connection with important political missions, such
as aerospace, defence and nuclear energy. When political priorities
change, as they tend to do fairly regularly, decisions about resource
allocations often change too, causing instability and confusion in the
longer-term pure and strategic research activities that depend on the
political missions for their support. In considering the special vulner-
ability of pure research in this situation, Brooks said:

> Public disenchantment with the military–space–atomic energy syndrome
> of the past quarter century has been visited on the fields of pure research
> that rode on its coat-tails for so long. When the only public function to
> which the electorate seemed willing to devote large resources with no
> questions asked was national security, it was natural to justify the support
> of science under this label. The magnitude and complexity of the defence
> budget, and the tendency of Congress to concentrate its attentions on the
> big hardware items, meant that research received little detailed scrutiny
> and this became a device for protecting the autonomy of science.[3]

So the political missions which provided the pure scientists with
unquestioned support for so long eventually caused these same
scientists' activities to be questioned. The need to justify basic research
in its own right became more pressing when public enthusiasm for the
great technological spectaculars began to wane. The third incentive
for developing well-defined criteria for allocating resources to R & D is
the recognition of the political reality, so starkly put by Alvin M.
Weinberg, that if those engaged in scientific research do not develop
such criteria and make hard choices, then 'these choices will be made
for them by the Congressional Appropriations Committees or the
Office of Management and Budget in the United States and similar
bodies in other countries.'[4] Finally, in the last two decades, governments have tended to demand more accountability in the area of
R & D expenditure, a trend that the OECD acknowledged in 1971
when it reported that even basic research would have to respond to
needs arising from the social, political and industrial environment.[5]

Decisions to allocate resources to R & D are made at three levels: the
gross national expenditure level (GNERD), the intersectoral level
(which, in most countries, corresponds loosely to divisions between
ministerial portfolios) and the intrasectoral level. The gross national
expenditure on R & D, which the scientific community ritually regards
as inadequate, is simply not amenable to any kind of analysis that
would set it relative to the levels of government expenditure in other

areas. In many countries, government officials have had to be content with adopting the rules-of-thumb mentioned earlier. At the peak of its investment in R & D, the United States reached a GNERD level of just over 3 per cent of the GNP. This level now stands at about 2 per cent, and this is the figure that scientists in countries that allot a lower percentage would like governments to recognize as a minimum standard. The proportion of the GNERD that should come from the public sector is generally not spelt out, but if it goes above 50 per cent this is usually regarded as a matter for concern, as the private sector should not be overly dependent upon government or overseas R & D.

The level of gross national expenditure on R & D, and the way in which it is distributed between the public and private sectors, have rarely been analysed. Attention in the literature has tended to focus on how given allocations are divided between basic, strategic and applied research and between those sectors of government that perform, commission or fund strategic and applied research. Despite the fact that most R & D funding in industrialized countries is for strategic or applied purposes, there has been a concentration in the literature on developing criteria for choice and identifying priorities in pure research. In this discussion, however, criteria for choice in basic (pure and strategic) research will be discussed as a whole, and it will be obvious when the considerations being put forward apply to pure research alone.

Basic research: the high civilization and overheads principles

In the early 1960s, when the debate on criteria for choice was in full swing in the pages of the prestigious science policy journal *Minerva*, Dr Alvin M. Weinberg, Director of the Oak Ridge National Laboratory in Tennessee, identified two 'principles' that justify public support for basic research. These were the high civilization principle and the overheads principle. The high civilization principle justifies basic research on the grounds that the search for new knowledge about humanity and nature is intrinsically valuable; it should be supported for its own sake in the same way that the arts are supported. Scientific knowledge is an indispensable element of human culture, and in view of its universal nature, those nations who fail to support it at a level commensurate with their wealth are neglecting their responsibilities to the international community and are parasites on the intellectual efforts and financial sacrifices of others. An economist who justified basic research in this way would define it as a consumption good.

The overheads principle looks upon basic research as a charge on society's entire technical enterprise. Strategic research should be funded by government as an overhead charge on the applied research to which it is most directly relevant. According to Weinberg, every good applied research laboratory should allocate a certain proportion of its funds to strategic, long-range research. Those who allocate the most, he claims, have the most success in achieving their missions, so it is only natural that expenditure on strategic research should be regarded as an overhead charge on an agency's mission. Pure research should be funded as an overhead charge on society's entire technical enterprise, because 'in a general and indirect sort of way it is expected eventually to contribute to the technological system as a whole.'[6] Thus, the overheads principle justifies basic research in economic terms—strategic research because of its relevance to the applied research for which it provides infrastructural information and problem-solving capability, and pure research because of its long-term usefulness. The economist Carl Kaysen, who also contributed to the *Minerva* debate, assumes, like Weinberg, that basic research of all kinds is justified economically. 'It is,' he says, 'a capital investment in the stock of knowledge which will pay off in increased output of goods and services that our society strongly desires.'[7]

However, neither the high civilization nor the overheads principle provides us with any criteria for deciding the overall level of expenditure. Weinberg leaves it to the mission-oriented agencies themselves to decide the amount that they devote to strategic research, and he leaves it to the politicians to decide the amount that should be allocated to pure research. So, as with gross expenditure on R & D, no criteria have been developed to assist government decision-makers to decide the proportion of the national research expenditure to be devoted to strategic and pure research. And as with the gross expenditure levels, the percentages decided on are usually based on international comparisons or rules-of-thumb. A figure of 10 per cent is generally quoted as the maximum proportion of pure research that should be allowed in the mission-oriented government agencies, but there is no evidence that this is the best, or the worst, percentage.[8]

The high civilization principle and the republic of science

Even though the high civilization and overheads principles offer no guidelines for deciding the overall level of support that should be given to pure and strategic research, do they suggest any criteria for allocating funds between areas of basic research? In Weinberg's scenario, the criteria for funding basic research would necessarily reflect those for funding the applied research the basic research is supposed to 'serve'. In the case of pure research, if this were to be justified by the high civilization principle alone, then no such criteria could possibly be applied. The best research would be supported for its own sake and to enhance the researchers' and the nation's reputation in the international community of scientific scholars, and these scholars alone would be capable of evaluating the potential contribution of a particular piece of research to the advancement of knowledge about humanity and nature. The public funding of pure research is not usually justified in terms of the high civilization principle alone. If it were, in all probability the advanced industrialized nations would spend only a fraction of what they currently spend on this research. Even the most eloquent advocates of substantial government support for pure research would hesitate to seek the current level of support solely on the grounds that pure research contributes to a nation's cultural heritage and embellishes its international reputation. Weinberg, for example, laments that

> until and unless our society acquires the sophistication needed to appreciate basic science adequately we can hardly expect to find in the admittedly lofty view of 'science as culture' a basis for support at a level which we scientists believe to be proper and in the best interests of both society and science.[9]

From the economist's point of view, Professor Harry Johnson equates the science as culture arguments with the arguments put forward in former times to justify a leisured existence for those who were fortunate enough to own land. He leaves us in no doubt that, in his opinion, scientific research pursued as an element of high culture would have very low priority in competition with society's pressing needs.[10]

Basic research, both pure and strategic, does in fact help to generate important technological innovations; precisely because of this, governments support it to a far greater extent than would be warranted by the high civilization principle. The scientific community has tended to capitalize on the technological successes of pure science to justify the levels of support that it now needs to satisfy the ever-increasing demands of 'big science'.

Despite the fact that the high civilization principle cannot be used to justify massive public support, the pure science community and to a certain extent those engaged in strategic research have tried hard to preserve the freedom of choice that this principle entails. Brooks has said:

> The frequently demonstrated usefulness of 'useless' research is sometimes used as an argument against any kind of social guidance of research. Like the invisible hand of the classical free market it is argued that the autonomous workings of the intellectual free market of ideas will produce socially optimal results at the least cost. The most eloquent exponent of this view was Michael Polanyi, who coined the term 'the Republic of Science' as a symbol of the idea of scientific autonomy.[11]

Polanyi, who was Professor of Chemistry and subsequently Professor of

Sociology at the University of Manchester, is most closely associated
with the term 'republic of science', but he did not coin it. The term is
found in Babbage's polemical work *On The Decline of Science in
England* written in 1830, and elements of the concept can be seen in
the writings of the various participants in the freedom/planning
debate that took place in Britain during the 1930s.

In 1933 Professor A.V. Hill, Nobel Prizewinning Professor of
Physiology at the University of Cambridge, drew attention to the fact
that for several centuries scientists had been regarded in all civilized
communities as entitled to a certain freedom from interference or persecution. The rationale for this freedom resided in the method of
scientific thought, the direct appeal to nature by experiment, and the
new powers given to humanity in general by scientific applications.
These characteristics, Hill claimed, do not obviously depend upon the
opinions, emotions or interests of any limited group, and so 'in a
certain sense science and learning are superior to and above the state.
All civilized people will admit that science transcends the ordinary
bounds of nationality.'[12] Thus, according to Hill, the international-
ism of science and the objectivity of its methods should protect
it from interference or direction by outside agencies, and this was
essential if science was to progress and its benefits were to be realized.
The scientists for their part must heed Hooke's dictum and avoid
meddling with 'Divinity, Metaphysics, Moralls, Politicks, Grammar,
Rhetorick, or Logick'. In an amusing reply to Hill's idealism, Professor
C.H. Waddington, one of a small group of eminent British Marxist
scientists who opposed this *laissez-faire* attitude towards science and
argued strongly that governments should direct scientific activities to
areas of national need, claimed that since Hooke's day science had
meddled in human affairs 'as gently as a Molotov cocktail in an
ammunition dump.'[13]

But it must be realized that Hill was thinking of pure science in the
universities when he formulated his opinions. He abhorred the
tendency then becoming apparent of attaching branches of science
and technology to industry and government departments, and sought
freedom of thought, freedom of communication, and freedom from
the sort of physical interference in scientific affairs that was occurring
at the time in Nazi Germany and the Soviet Union. It is not difficult to
see why he linked the universal character of pure science with these
particular freedoms, but it is more difficult to discern his reason for
claiming that because of its very nature and its international character,
pure science is above the state and that the new powers it gives to
human beings are necessarily independent of the interests of what he
called 'limited external groups'.

Another prominent member of the anti-planning group of the
1930s, Dr J.R. Baker, a lecturer in botany at Oxford University who,

with A. G. Tansley and Michael Polanyi, formed the Society for Freedom in Science in response to the planning movement, considered the freedoms demanded by Hill to be very small concessions indeed. 'Many people,' he said, 'who talk airily about the necessity for freedom in science are actually thinking of the freedom of scientists to say and do as they like. This is a relatively small freedom if they are not allowed to investigate what they like and, therefore, make discoveries which are worth talking and writing about.'[14] Baker traced the origins of the iniquitous movement towards planning in science to Francis Bacon — 'a less suitable object for heroics can scarcely be conceived' — and of all of Bacon's words only a few found favour with this defender of the pure science community's autonomy. Predictably, these were the passages in *New Atlantis* that described the work of the pioneers — 'those of whom do try experiments such as themselves think good'. Baker believed that science could not be planned because its results are unpredictable. Writing before the large research teams of 'big science' had appeared, he thought that creativity could not be nurtured in a team and that discoveries were made as a result of serendipity and not by plan. Like Hill, Baker concerned himself with pure science in the universities. He conceded that it would not be unreasonable to require researchers in specialized research units to work in areas that were relevant to the unit's mission. And, demonstrating that even in the case of these special research units he was thinking of those in universities and not in industry or government, he demanded that under all circumstances investigators should be free to publish their results, whatever these might be. Thus Hill, in his own way, and Baker, more explicitly, demanded pure scientists' right to choose freely the problems they would work on. In the 1930s, when the level of public support for pure science was so low that it could easily be justified in terms of the high civilization principle, this demand was by no means unreasonable.

Michael Polanyi's views on freedom in science have been influential, and scientists' present day demands for autonomy in choosing areas of research and distributing resources between areas are based on his notion of 'the republic of science'. Polanyi's central thesis is that the scientific community is a society of explorers striving to discover truths about the natural world for intellectual satisfaction. Choosing their own problems and judging for themselves what is feasible and worth doing, scientists are members of a closely knit group whose activities are co-ordinated by individual members adjusting their efforts to the existing results of others. In this they are guided by an 'invisible hand' towards the discovery of the hidden system of things.[15] The scientific community is similar, therefore, to a body politic and works according to a system that has much in common with economic *laissez-faire*, where the invisible hand of the market system guides entrepreneurs to profitable areas of economic activity. Any attempt by external authorities to interfere with the workings of this republic of science would, in the end, be self-defeating. Polanyi himself puts it like this:

> I appreciate the generous sentiments which actuate the aspiration of guiding the progress of science into socially beneficient channels, but I hold its aim to be impossible and nonsensical...Any authority which would undertake to direct the work of the scientist centrally would bring the progress of science to a standstill.[16]

It can be inferred from Polanyi's analysis that no criteria can be developed to guide decision-makers in distributing resources to pure science, apart from the quality of the investigators and the scientific merit of their research proposals. These judgments can be made only by scientists. This autonomy is demanded, not as an inalienable right but because it makes the research system as a whole more efficient: if the progress of science is slowed or brought to a halt by external interference, then the benefits that flow to society, too, will be delayed or halted. Polanyi assumed, of course, that social benefits would indeed flow from pure research activities.

Although the concept of an autonomous republic of science is popular in the scientific community, it has also been severely criticized. Brooks has dismissed it as 'scarcely viable in a polity in which a large fraction of the resources invested in the enterprise of science comes from the public and in which the total investment is as large as it is.'[17] A special Canadian Senate committee on science policy (the Lamontagne Committee), which reviewed the literature on science policy in the course of its searching examination of the nation's research system, rejected the republic of science as an organizational strategy for a national R & D system on three grounds. First, the concept has the same origins as economic liberalism, and since it is now recognized that society cannot rely on the 'independent self-coordinated initiatives' of private producers to maximize social and economic progress, it is unlikely that it will be able to do so in the case of the producers of scientific research. Second, one of the important requirements of economic liberalism is that private producers use their own funds to pursue their objectives. Finally, in the Committee's view, the major scientific and technological developments since the Second World War that have been guided by social and economic objectives contradict Polanyi's statement that attempts to direct scientists' work would bring scientific progress to a standstill.[18]

A major criticism of Polanyi's approach is that it fails to appreciate the political context in which science is now practised. Polanyi believed that the public was not interested in science as a source of wealth and power alone. This is almost certainly true, but it is unlikely that the public would be willing to allocate the vast sums that are now given over to basic research if the republic of science dictated the mechanisms of research allocation, co-ordination and control. It is important to emphasize, however, that Polanyi was concerned only with pure research in universities. He made a clear distinction throughout his work between science and techñology, between pure science and applied science, and between those who were engaged in the pursuit of scientific knowledge for its own sake and those in industry and in government service who pursued it for specific aims. He did not extend to the non-university sector the autonomy that he considered necessary for the pursuit of pure research. 'No national pursuit,' he said, 'can be guided wholly by its accidental results. However close the symbiosis of science and technology may be, each forms a separate organization for which its own vital interest must serve as its guide.'[19] Pure science does not have the same object as strategic and applied science, and the concept of the republic of science does not therefore apply to the latter.

The distinction that Polanyi made between pure and applied research does not seem to have been readily appreciated by the scientific community, judging by the writings of some leading scientists. Writing in the wake of the publication of the first volume of the Lamontagne Committee's report, *A Science Policy for Canada*, Dr Gerhard Herzberg, Distinguished Research Scientist at the National Research Council of Canada and Nobel Prizewinner in chemistry, perceived the republic of science to be in mortal danger at the hands of the inquiring senators who were calling for a coherent science policy for Canada.[20] In the tradition set by J.R. Baker, Dr Herzberg identified himself with a rapidly disappearing breed — a scientist who is interested in doing science and not in controlling those who do it — and, like Baker, used a military analogy to describe the rise of the science bureaucracy: four officers for every private was his estimate! Setting out his own definition of science policy, derived from that proposed by the British Council for Scientific Policy and unashamedly 'policy for science',[21] Dr Herzberg feared that this was not quite what the senators had in mind when they called for a coherent science policy. What they did have in mind, he thought, was a scientific plan, formulated centrally and directed by a Minister of Science and departmental staff.

A great deal of the Canadian Government's research effort is carried out in Crown Corporations, established by law and reporting directly to a minister rather than to the permanent head of an administrative Department of State. The National Research Council (NRC) is an illustrious Crown Corporation, and the concept of the republic of science is deeply embedded in its scientists' consciousness. Dr E.W.R. Steacie, its distinguished second Director, expressed the NRC's dominant principle of management when he announced that the fundamental feature of the administration of Canadian science in the Crown Corporations would be to make sure that administrators could never issue any instruction to scientists on any technical matter whatever. In Dr Herzberg's view, the republic of science Dr Steacie advocated was endangered by the Senate Committee's activities and recommendations; freedom of thought, communication and choice of problems were threatened. It is important to note that Dr Herzberg was an NRC scientist—a very distinguished one—but nevertheless a Canadian Government employee carrying out research in a government scientific organization devoted to the pursuit of strategic and applied research. The republic of science that he considered to be in danger in the NRC was inappropriate as a guide for directing the NRC's work.

Some years later, another senior NRC scientist, Dr J.D. Babbitt, conducted a campaign against the establishment of any kind of coherent science policy in Canada, which he saw as the collectivization and socialization of research. Babbitt considered the Canadian scientific enterprise to be under attack from two forces, the anti-rationalist forces of the counterculture and the less fanatical but insistent ideology of science policy which sought to organize, systematize and direct all research for specific social goals. This pernicious ideology, he said, ran counter to the liberal individualism that for so long formed the philosophical background of the republic of science.[22] Babbitt, too, was extending the principles governing the republic of science to the government sector. Like Herzberg, he saw the Lamontagne Committee's report as a threat to the NRC's traditional autonomy, which had been conferred on it by its corporate status. Many Canadian scientists believe this status to be the organizational solution to the problems faced by government scientists, who, it is claimed, need to feel autonomous in choosing problems, methods and communication in order to be fully creative and productive.

In Australia, too, corporate status for government scientists is widely believed to be an appropriate form of organization, but with Australia's entry into the debate a new element was introduced: the proposition that all government scientists, not just those engaged in long-range strategic research, should be accorded the autonomy that is appropriate in the republic of science. The arguments are set out in a paper in *Minerva* which summarized the findings of a special science task force, composed mainly of distinguished scientists, appointed in 1975 by a Royal Commission into Australian Government Administration.[23] Taking creative, productive science as the desirable end product of the government's intramural scientific effort, the science task force, led by Dr J.R. Philip, then Chief of the CSIRO Division of Environmental Mechanics, attempted to define the conditions under which such a goal might be attained.

The task force claimed that creative, productive science is possible only if the four norms of science identified by Robert Merton are preserved.[24] According to Merton, these norms are an essential component of the scientific ethos, 'that effectively toned complex of values and norms that are held to be binding on the man of science'. They are both desirable and necessary as the means by which certified knowledge is extended. *Universalism*, a fundamental precept, requires that universal criteria are used to evaluate contributions to the body of scientific knowledge. *Communism* requires that scientific results are shared within the scientific community. *Disinterestedness* on the part of scientists is an institutional mandate. It is fostered among individual scientists as long as their work is under the exacting scrutiny of fellow experts but is at risk once scientists are judged by other standards. *Organized scepticism*, an institutional and methodological mandate, requires that judgment be suspended until the facts are at hand. Prestige and authority should play no part in the evaluation of scientific contributions. 'The borrowed authority of science should not bestow prestige upon the unscientific doctrine; claims to truth are based on the testable character of science, not on non-scientific criteria.' These standards are preserved only if the scientific community remains autonomous. The task force's inevitable conclusion, then, was that scientists could be creative and productive only within an organizational structure that allowed them almost complete autonomy. In general, the statutory or corporate form of organization, as found in the CSIRO and NRC, was considered to be the one that came closest to this ideal for government scientists, and it was recommended that this form be extended as far down the hierarchy as possible to almost the whole of the Australian Government's R & D system.

If the Herzberg–Babbitt essays took the republic of science out of academia into the strategic research sector of government, the Australian science task force took it one step further. Not until then had it ever been suggested that autonomy was a necessary condition for the conduct of creative, productive science of all kinds—pure, strategic and applied. The implications of this view are quite far reaching. If we separate 'autonomy' into two components—professional and operational—we shall see just what it is that was being proposed. In a democratic society, no one would seriously question the proposition that scientists should be autonomous in evaluating research proposals, validating research data, judging techniques of investigation and conferring professional recognition and esteem. This we shall call professional autonomy, a clear prerequisite if scientists are to work effectively. In demanding autonomy of this sort, which A.V. Hill and Michael Polanyi were doing, scientists are pressing a claim that should be granted to any professional group.

However, like J.R. Baker, the science task force must have considered this to be a small privilege. If professional autonomy were all that were sought, then it would hardly be sufficient to extend it to as low a level as possible in the scientific community—it is a legitimate and uncontroversial claim at all levels. The task force's thinking on the matter is revealed in the following passage:

> Traditionally, the work of science has been conducted essentially by an autonomous self-directed community. Decisions about research directions, judgements about the validity of data or theories, recognition of achievement and granting of esteem have been in the hands of the scientific community itself. It is well known that this traditional autonomy of science is being increasingly called into question.[25]

Thus it is not just professional autonomy that is being sought but also freedom to set the agenda for research; freedom to decide not only how problems might be solved—which is readily accepted—but also to decide which problems will be tackled. This can be called operational autonomy. While it is accepted that there is a necessary connection between professional autonomy and creative, productive science, no necessary connection has ever been demonstrated between creative, productive science and operational autonomy. Indeed, if creative, productive science is possible only in an organization that allows full operational autonomy, then the majority of scientists are not in the best working environment. Only in the universities and other institutions of higher learning is such an environment to be found, at least in theory, and it is not true that it is only in such institutions that creative, productive science takes place (as two examples given in Chapter 2 show). The key scientists engaged on the Manhattan Project were working in a closely controlled environment in which they had full professional autonomy but did not decide which problems they would tackle. In the case of Interscan, too, it is clear that the CSIRO research scientists were no less creative and productive for not themselves choosing the problem they tackled.

Summarizing at this point, it may be said that if public support of basic research is justified in terms of the high civilization principle, then a self-governing scientific community, selecting projects that will advance scientific knowledge and distributing funds on the basis of scientific merit, would be the most appropriate and effective form of organization. In fact, the high civilization principle is now rarely used as the sole justification for public support, because it is realized that less support would be given on these terms. Despite this, the scientific community has consistently lobbied for the resource allocation mechanism and organizational structure appropriate to a republic of science operating on the high civilization principle. Some have argued that this form of organization should be extended to strategic and even applied research, even though this was not advocated by Polanyi, the originator of the republic of science concept, and it has by no means been proved that working conditions appropriate to the pursuit of pure research as an element of high culture are equally appropriate to mission-oriented research.

The overheads principle and Weinberg's criteria

As we have seen, the overheads principle offers an economic rationale for public support of basic research: it is expected that society will eventually reap the benefits of the investment in terms of goods and services. Despite the difficulties of establishing a direct connection between basic research and technological innovation, economists have assumed a direct link and have attempted to formulate qualitative criteria for establishing priorities.

In the 1950s C.F. Carter argued that scientific effort should be channelled into areas that will maximize the flow of material wealth. At that time export performance was a pressing problem in the British economy, and Carter suggested that part of the national research effort might well be directed to forecasting the pattern of exports ten or twenty years into the future. Thus:

> By asking these questions about long term prospects of British exports it would be possible to reach some conclusions, first about applied research and development, then about the pure research which feeds into it and finally about the forms of training needed to support the research.[26]

In Carter's strictly economic view, a nation that funded basic research to a greater extent than could be justified economically was placing more importance on the support of science as culture than it was on increasing wealth. Professor Harry Johnson echoed this opinion when he said that if the public is convinced of the value of basic research as an activity with no economic function, then it is appropriate that this research should be paid for out of the public purse. Both economists tried to place the allocation of resources to basic research in a wider social context, but the central problem still remained: that is, what amount of basic research is justified economically? There was no purely economic answer to this question, but if basic research were regarded as an overhead charge, either on applied research or on society's total R & D expenditure, it might be possible to derive criteria for allocating funds between areas of research.

Kaysen, as noted, viewed expenditure on basic research as an investment in a special kind of social overhead — knowledge and understanding — that contributes both directly and indirectly to a wide variety of social purposes. Basic research, he said, makes a major contribution to the advancement of applied science and technology and has helped to train specialist scientists and engineers — experience shows that R & D programs aimed at solving specific practical problems benefit from a close association with scientists engaged in basic research. Finally, the corps of scientists engaged in basic research represents an important reserve of capability that can be drawn upon in times of national emergency. Kaysen recognized, however, that these considerations do not help us to decide what level of expenditure on basic research is either necessary or desirable, and he was sceptical of any attempt to derive any such criteria based on return on investment. With regard to allocations between fields, he was forced to admit that 'in the absence of an objective standard for judging, say, that particle physics ought to be developed faster than radio-astronomy and in the presence of budgetary constraint, the allocational decision must inevitably represent somebody's preferences or prejudices.'[27] The problem was still there: no objective criteria to assist in the allocation of resources had yet been formulated. Weinberg, however, offered some criteria for choice based on the overheads principle that would, he believed, deliver the policy-makers from the impasse.

Weinberg developed a set of qualitative criteria, which he divided into two types — extrinsic and intrinsic. Intrinsic criteria should be applied within the scientific field itself and answer the question: how well is the basic research likely to be carried out? Extrinsic criteria are generated outside the field and answer the question: why pursue this particular piece of basic research? Two intrinsic criteria can be readily identified. They are:

• Is the field ready for exploitation?
• Are the scientists in the field competent?

Of the two sets of criteria, Weinberg considered the extrinsic ones to be the more important. There are three extrinsic criteria, relating to the technological, scientific and social merit of the proposed research. Technological merit is fairly straightforward and reflects society's desire, through its elected representatives, to achieve a particular technological advance. Once that decision has been taken, the necessary R & D has to be supported. The broader question whether the technological aim itself is worthwhile has to be answered partly from within the technological area itself and partly from outside it, while the questions whether the field is ready to be exploited and whether the workers are competent must clearly be answered from within the area. The question whether the social objectives that the technology is designed to achieve are desirable must rest with the wider society.

The criterion of scientific merit is more difficult to apply than that of technological merit, and it cannot be judged entirely from within the field in question. According to Weinberg, the only valid measure of scientific merit is relevance to neighbouring fields of science. The field that contributes most to neighbouring scientific disciplines has the most scientific merit.

The intrinsic criteria and the criterion of scientific merit embody a recognition of scientists' need for professional autonomy. They operate through a peer review process, which is based on the assumption that there are more or less universal criteria, on which any group of competent scientists will agree, for judging proposals. In recent years, there has been some controversy over the merits of peer review, not so much concerning scientists' ability to evaluate their colleagues and rank scientific projects in order of priority, but concerning its appropriateness as a means of formulating science policy and allocating resources between fields. Brooks has suggested that peer review operates most successfully at the truth rather than the utility end of the scientific spectrum; that is, peer review is more successful in pure than in applied research. According to him, the failure of the United States National Science Foundation's Research Applied to National Needs (RANN) program was due in no small measure to the inadequacy of peer review of applied research.[28] It may be, he says, that at the applied end of the spectrum the system may be manipulated by unscrupulous 'entrepreneurs' who by giving their proposed programs useful-sounding labels make them sound more attractive to politicans and avoid rigorous scrutiny.

Under the peer review system, the number of proposals in various areas is supposed to indicate to program and fund administrators what the scientific community considers to be most worth doing. But Brooks sees some problems arising from this system. First, it can be distorted by the availability of funds. Many researchers will generate proposals in the areas where funds are known to be available, irrespective of the intrinsic scientific or social value of the proposed research. Second, the number of proposals in a particular area may reflect earlier support for the field, which, with the personal, institutional and even bureaucratic commitment that this entails, leads to self-perpetuation rather than recognition of new opportunities. Third, peer review is a poor indicator of over-commitment to a field, especially in a pluralistic situation as exists in the United States, where many agencies are responsible for funding basic research. Proposals are usually received at different times by different agencies and reviewed by different referees, and so over-concentration of resources in a narrow field will not be readily apparent. Fourth, proposal pressure tends to favour safe rather than highly original projects that do not fit accepted paradigms — it is essentially a conservative procedure. Finally, it is difficult to compare projects under this system. The number of research workers in different fields of scientific endeavour varies greatly, and the 'pressure' coming from a sparsely populated field may have quite a different significance for that field's needs than that coming from a mature field that historically has been well endowed.

Because of these criticisms, proposal pressure is not a good indicator of grass roots priorities in basic research and cannot be used as a major criterion for allocating resources. It is, however, an important signalling device, which scientific administrators cannot ignore when judging scientific merit. Recognizing the limitations of peer review in this context, Martin and Irvine, in a series of pioneering and controversial papers, have combined it with publication counts and citation analyses to assess the performance of a number of European research groups and institutions.[29]

Weinberg's third criterion — social merit — is the most difficult of all to apply. Two difficulties are immediately apparent. First, who is to identify a given society's values, and second, even if a social value can be identified satisfactorily, how can it be decided that a particular scientific or technological enterprise will further the pursuit of that value? Weinberg himself points to the difficulty of deciding which scientific or technological activities would enhance national prestige and, having identified these, of then setting priorities. For example, does high-level basic research leading to national leadership in the Nobel Prize stakes rank higher than discovering a vaccine to combat polio? Does landing someone on the Moon outrank international co-operation through science?

Weinberg's criteria certainly make the problem of choice more explicit, and despite their qualitative nature and the problems associated with the criterion of social merit, they have had an enormous impact in the science policy arena. Their influence can be seen in the United States in publications of the National Science Foundation, the National Institutes of Health and the National Academy of Sciences. In Canada, the Science Council of Canada has in the past been heavily influenced by Weinberg, and in the United Kingdom, Gibbons saw the criteria being applied when the Council for Science Policy recommended that the British Government join the European co-operative high energy physics consortium, CERN, based in Geneva.[30]

In the United States, a survey committee of the National Academy of Sciences applied the criteria to physics research in a way that was both sophisticated and influential. Twenty-three criteria in three different categories were compiled. Physics was divided into a number of subfields, and these in turn were divided into program elements. The Academy's physics survey committee was then asked to vote on each of the program elements set against each of Weinberg's criteria. It was found that the program elements could be divided into two groups — those with high intellectual merit and those with high social merit. Some, like quantum electronics, scored high in both of these categories, but this outcome was rare.

The committee was unable to resolve the basic difficulty of how to weight the various criteria. The judgments made were highly subjective and eventually a mixed strategy for the development of physics was recommended — some program elements with high intellectual merit and some with high social merit. In other words, it proved impossible to rank projects purely by applying the criteria. Despite the fact that the committee stated explicity that it did not feel that its judgments were good enough to be acted upon, there is no doubt that the recommendations were heeded. According to Brooks, it is probable that the level of funding for basic research in physics decreased less rapidly than it otherwise would have, and certain recommendations in astronomy and high energy physics were followed.

To sum up, public funding of basic research as defined here can be justified in terms of high civilization or as an overhead on other scientific or technological missions. Though both pure and strategic research can be justified in terms of high civilization, in practice only pure research is usually justified on this ground as well as economic grounds. The organizational principles embodied in Polanyi's republic of science require that resources be allocated to pure science carried out as an element of high culture on the basis of scientific merit as judged by the scientific community itself, and there seems to be a general rule-of-thumb that this sort of research should command about 10 per cent of a nation's gross expenditure on R & D. Though the scientific community generally considers that resources should be allocated to strategic research on the same principle, the autonomy that this would give scientists engaged in mission-oriented research would be considered difficult to justify both politically and economically. Weinberg's criteria, based on the overheads principle, provide a qualitative means of establishing priorities among basic (pure and strategic) research projects in the mission-oriented agencies of government once these agencies have been allotted their funds. But no criteria, qualitative or quantitative, have yet been developed which might help governments to decide how best to distribute R & D funds between the various sectors represented by ministerial portfolios.

Applied research in government agencies: science policy machinery

It is surprising to find that, despite the fact that most research funded by the governments of the OECD countries is either carried out in their own mission-oriented laboratories or contracted to outside agencies, little attention has been paid in the science policy literature to developing criteria for allocating resources to priority areas in the different sectors of government that are responsible for spending such vast sums on R & D. There is, it seems, an implicit assumption that applied research in the departments and mission-oriented agencies of government will reach an appropriate level determined by competition with other government activities and response to need. As Brooks puts it, this implies that a nation's social priorities are identified by the political process and the research system is deployed to tackle the problems posed in the most effective manner. Research priorities are decided as a result of a constant dialogue between the political process and the government research system, and the appropriate organizations then ensure that priorities are identified and research activities co-ordinated.[31] As a result of this thinking, science policy research into criteria for choosing projects and setting priorities in government research agencies has focused on setting up appropriate forms of organization to ease the problem of choice and maximize the efficiency of government R & D efforts.

Governments support research activities in their departments and agencies because they are convinced of their social, economic and political value. The organizational systems that they adopt from time to time to co-ordinate and control these activities reflect their current views on the relationship between scientists and society at large. It was stated in Chapter 1 that the turning point in the relationship between science and government came during the Second World War, when scientists showed that with investment and organization they could achieve unprecedented results. From then on scientific research was seen as a vital national resource, and new organizational structures were needed to cope with the new breed of 'scientific' civil servants who would enter public administration with requirements and demands quite different from those commonly regarded as acceptable in the public service.

The novelty of the situation accounts for the improvization apparent during the post-war years when the idea of a national science policy was taking hold. As Salomon observes, the wartime arrangements were used as a model, and tensions were inevitable. On the one hand, there was the pluralistic tradition of science that acknowledged diversity of disciplines, institutions, funding arrangements and approaches to problems, and on the other, the forces of government that stressed accountability, lack of duplication, planning and centralized control.[32] In wartime the overriding objective of survival tended to diminish the tensions created by these different styles and preferences, reducing them to a manageable level so that the entire partnership was not placed in jeopardy. In peacetime there was no such overriding factor, and the pre-war freedom/planning debate emerged in new forms.

It has already been pointed out that in government science, competition for R & D resources has encouraged both scientists and administrators to consider more 'rational' or scientific means of allocating resources. Priority setting could be envisaged in a rational decision-making process. First, priority areas would be identified by some means. Then, in the light of these priority areas, new initiatives and existing programs would be assessed. Finally, resources would be allocated to the chosen programs. The essential backcloth to all this would be a set of national objectives to which R & D could reasonably be expected to contribute and an organizational structure that enabled these needs to be translated into research programs. The problem of identifying national objectives will be considered later on in this chapter when analytical methods in science policy are discussed.

The Senate Committee on Science Policy in Canada and the Dutch Ministry of Science Policy have identified four types of science policy organizations, which they define as pluralist, co-ordination, concerted action and centralist.[33] Brickman, in a later analysis, reduced them to three — atomistic, low co-ordination and high co-ordination,[34] but in this review the original four will be used, with references to Brickman's three.

THE PLURALIST SYSTEM

Pluralism implies diversity, and in a pluralist system financial resources are assigned to each policy sector as a whole. Government departments receive an appropriation at the beginning of the fiscal year, and it is up to each department to decide, by whatever means it considers appropriate, the amount of money it will give to its R & D activities in toto and then to the individual elements within the portfolios. Each department operates independently and usually sets its own priorities. Apart from some overall supervision by the Department of Finance or Treasury, the national science policy is the sum total of the activities undertaken in the departments under the general authority of the Minister and executed in most cases through the Permanent Head and the civil service staff.

Brickman describes as atomistic the situation where there is no explicit central framework within which the actions of an individual department or quasi-autonomous agency can be judged in relation to those of other departments or agencies. Each element, he suggests, operates independently, manoeuvring to achieve its objectives in the light of how it perceives the others' actions. The advantages of pluralism are that the departments are encouraged to develop their own expertise in research management and policy; decisions can, or should, be taken fairly quickly within each department; and, it is claimed, this system offers the best guarantee that research results will be applied in practice. Schmandt, who considers pluralism to be one of the central principles of science policy, claims that the atomistic system offers protection against errors of judgment in resource allocation decisions. According to him, uncertainties about the outcome of proposed research programs cannot be eliminated from such decisions, and a strategy of attacking problems from different points of view reduces the risk of missing a breakthrough, a risk always attached to the more single-minded approach adopted by some other systems.[35]

However, to define pluralism as a science policy 'constant', an enduring principle which should be kept in mind when decisions are being made on funding, organization and linkage to other policy goals, is unwarranted. In certain circumstances this form of organization can be extremely effective. In a very wealthy nation, for example, where there is not a great deal of pressure on resources for R & D and it is not vital to develop criteria for choosing and setting priorities among research projects, an uncoordinated, pluralist system of allotting funds may well be socially, politically and economically desirable. In a national emergency, multiple approaches to a vital problem may, indeed, be essential. The decision to develop simultaneously the three possible methods of isotope separation in the Manhattan Project was taken because the United States Government could afford the expense and the participating nations' survival was considered to be at stake. In peacetime, however, and under conditions of financial restraint, duplication of this sort is difficult to justify. This applies *a fortiori* to much pure research, where the problems are rarely so important as to warrant multiple approaches to solving them. Under these circumstances, a more co-ordinated approach to science policy would seem to be more appropriate.

THE CO-ORDINATED SYSTEM

In a co-ordinated system, the various government departments initiate projects independently, but there is a greater awareness of the need to co-ordinate their research activities in order to avoid duplication and to apply the results in practice. This system provides for central advisory bodies to form the links between the various government sectors responsible for R & D. These advisory bodies may be composed of eminent personalities from outside the government system, senior representatives of government departments or both. The co-ordinated system of organization would be included in Brickman's category of atomistic co-ordination.

The efficiency of the co-ordinated system is now in question after a number of years' experience in many of the OECD countries. As was pointed out in Chapter 1, one of the first manifestations within the OECD countries of an intention to establish a science policy machinery was the establishment of an advisory body of prominent scientists, industrialists and eminent citizens to advise the government on all matters relating to the development of the national scientific effort. In Britain, the Advisory Council for Scientific Policy (ACSP) came into being in 1947. Ten years later, the United States established the President's Science Advisory Committee (PSAC). European countries quickly followed suit with similar bodies, and Canada established its Science Council in 1966. In each case, after an initial period in which they played an important and influential part in their government's scientific affairs, the advisory councils went into a long period of decline. PSAC was abolished in 1973, and ACSP disappeared in 1964, to be replaced by a Council for Scientific Policy with greatly reduced terms of reference.

Many of the councils were victims of their own success. They lobbied consistently for coherent science policies and the establishment of ministries with responsibility for science or science policy. When these ministries and their associated bureaucracies were set up, the councils were inevitably moved to the sidelines. In France, for example, the influence of the Consultative Council for Research in Science and Technology (CCRST) was undermined by the creation of the General Delegation for Scientific and Technical Research (DGRST), a body that has now become a full Ministry for Research and Industry. When a Minister for Science Policy was appointed in The Netherlands in 1973, the influence of the Advisory Council for Science Policy (RAWB) waned, and in Belgium the National Council for Science Policy (CNPS) suffered a similar fate when a Ministry of Science Policy and an associated organization, the Science Policy Programming Service, were established. In Canada, the Science Council's influence declined when the Ministry of State for Science and Technology was set up in 1970, and in the Federal Republic of Germany, too, a Ministry of Research and Technology (BMFT) took over many of the Science Council's responsibilities.

Some of these situations will be examined in Chapters 4 to 8, when the science policy machineries of various OECD countries will be described, but from the examples that have been given it will be evident that the co-ordinated system, where external advisory bodies exercise loose control, has had limited durability. The institutionaliza-tion of science policy within the various bureaucracies was not the only reason for the decline in the external advisory bodies' influence,[36] but it was certainly one of the most important. With this institutionalization, new systems of organization were introduced.

THE CONCERTED ACTION SYSTEM

In the concerted action system, the government's research effort is co-ordinated by a Ministry, usually a Ministry of Science Policy. The Minister of Science Policy acts with the concurrence of all ministers with R & D responsibilities and is usually provided with a secretariat, which prepares background papers, provides budgetary analyses and formulates strategies. An interministerial committee finalizes the decisions that have previously been 'prepared' within the Ministry of Science Policy. This procedure, it is claimed, safeguards the interests of the different policy sectors and ensures that science policy looks beyond the boundaries that define the various government departments. The Minister of Science Policy assesses the R & D programs drawn up by the other ministries and administers a 'science budget', an *ex ante* compilation of all departments' proposed expenditure on R & D. Decisions on medium-term research expenditure are more easily made when the proposals of all departments and their agencies can be considered. In its extensive review of science policy mechanisms, the Canadian Senate Special Committee on Science Policy (the Lamontagne Committee) considered concerted action to be the best system, because it synthesized the main advantages of the other systems and had none of their shortcomings. This conclusion was based on the following criteria:

• a diffuse government organization for science, technology and innovation must be maintained;
• a strong and effective central machinery is needed to complement the specific science policies developed by operating and supporting departments and agencies;
• the central machinery's mandate must include specific authority to review and approve the science budget within the broad lines determined by the Treasury.

Though the Committee considered the concerted action model to have strong central machinery, this model is not a centralized system as defined below. In Brickman's nomenclature, the concerted action system would probably be defined as low co-ordination.

THE CENTRALIST SYSTEM

There is some confusion in the literature as to what constitutes a centralist system of science policy. For Brooks it means that all government scientific activities are administered by one Department of State with its own organization and funds. The Lamontagne Committee defines it as a system in which most of the government's R & D activities are concentrated in one Department of State, the other departments concerning themselves with short-term tactical research only—the British system between 1947 and 1964 was described as centralist by Lamontagne. These definitions are not entirely satisfactory; in this discussion, a centralist system in the context of the OECD countries refers to the type of system that has been in operation in France for many years.

The R & D system is placed within the framework of a national five year plan, and the various ministries' research programs should, in theory, be in harmony with the plan's objectives. This is one of the factors that distinguishes the centralist from the concerted action system. As in those countries with a concerted action system, proposed R & D expenditures are gathered together into a science budget (called a 'research envelope'), and the requests are sent to the Ministry of Research and Industry, whose Minister negotiates for the amounts involved with the Ministry of Finances and, on occasion, the Prime Minister. Once a decision has been reached, the agreed amounts are allocated directly to the various ministries, which put the approved programs into effect.

Generally, the logic behind centralization of the sort that operates in France and to some extent in Belgium is that most government research activities and scientific support programs are as homogeneous as other government responsibilities and can be organized and planned in much the same way. The advantages claimed for this system are that R & D objectives are integrated with those of other government departments and that priorities can therefore be identified and set more easily. Activities can be co-ordinated, the ways in which departments and agencies can co-operate are identified and it is easier to establish inter-departmental and inter-agency multidisciplinary programs. Working within a five year national plan means that funding is relatively stable for this period. It is claimed, however, that the centralized system leads to conservatism, rigid bureaucratic control of research and too great a separation between those who are doing the research and those who eventually make use of the results.

The Lamontagne Committee reached the following important conclusions about the pluralist, co-ordination and centralist systems. In the pluralist system, the diffuse decision-making process in individual government departments and agencies under the sole control of the Treasury is seen to be clumsy and inefficient as it becomes increasingly evident that science and engineering are inter-dependent and that multidisciplinary and inter-agency efforts are required to achieve economic and social goals. The co-ordination system has not worked in practice, because the advisory machinery, no matter how strong and complex it may be, has not significantly influenced government decisions. The centralist system isolates government R & D programs from the missions they are supposed to serve.

Setting priorities: rationalist and incrementalist approaches

A number of normative and empirical theories of public administration have emerged which attempt to throw light on the problems of how to make the best use of public resources in the pursuit of social and political objectives. The two theories that have been generally discussed within the context of science policy are the rationalist and incrementalist theories.

The rationalist approach to decision-making is based on the notion that the more informed decision-makers are, the better their decisions will be. Rationalist decision-makers identify, scrutinize and order the objectives and other values they believe should govern the choice of a possible solution. They then survey the various means of achieving these objectives and values and examine all the probable consequences of employing each of the possible means. Finally, they choose a means — a particular policy or combination of policies that will achieve what is required in an acceptable way. Rationality of this sort — where goals and objectives are set out clearly and all policy options considered — should not be pursued beyond the point of diminishing returns. In the end, decision-makers may have to be content with the first alternative they find which satisfies their basic requirements.

An alternative approach to management and decision-making is incrementalism, a procedure that is claimed to describe the way in which decision-makers actually work and, indeed, should work.[37] Given that no one can know everything there is to know in a particular area, decision-makers should focus on policies that differ incrementally from existing ones rather than attempt to survey and evaluate all the choices; they should then consider only a small number of policy choices; and finally, they should evaluate a restricted number of important consequences for each policy choice. The incrementalist approach allows for countless ends–means and means–ends adjustments, which make problems more manageable. Decision-makers proceed incrementally and sequentially, all the time taking account of objectives and the means of achieving them. By this process of step-by-step adaptation, piecemeal experimentation, bargaining and mutual adjustment between rival interests, decision-making is considered to become more democratic and effective. Proponents of the incrementalist approach do not favour overall co-ordination of activities or forward planning of the sort that is represented by, say, a science budget, because this thwarts the fundamental principle of allowing for ends–means and means–ends adjustments.

These two approaches to decision-making have advantages and disadvantages. The incrementalist approach may be valid if the results of current policies are, in the main, satisfactory and if the problems to be solved and the available means of solving them are not changing in kind. The approach can be criticized because it can foster inertia and conservatism and discourage innovation among administrators.[38] The rational action approach can be criticized, in turn, because it separates ends and means. This is undesirable, because ends and means must be considered together when deciding whether one policy choice is better or worse than another. Social interaction and conflict may be ignored if decision-makers focus on 'best way' solutions rather than on compromise and accommodation. Moreover, critics argue that there can be no single set of unambiguous, unchanging objectives with agreed priorities and means of implementing them.

The rationalist and incrementalist approaches to decision-making are paralleled in the broad organizational frameworks that governments have established to implement their science policies. Indeed, the frameworks themselves may reflect the dominant form of decision-making in the R & D area. A centralized or concerted action science policy machinery may be symptomatic of a rationalist approach, whereas the pluralist and co-ordination systems may reflect a more incrementalist approach.

Though the pluralist and co-ordinated systems of organization and the incrementalist approach to decision-making have many adherents, and are especially favoured by the scientific community, developments in many of the OECD countries make these approaches seem less appropriate or acceptable, and there is a trend away from the more *laissez-faire* systems. In particular, there is a trend towards establishing major national programs, involving many R & D sectors, aimed at solving pressing cross-disciplinary, interdepartmental problems. These national R & D programs are relatively recent phenomena, being a response to the so-called 'new economic and social context' that arose from the 1973 oil crisis, and they do not lend themselves to an incrementalist approach to decision-making or a *laissez-faire* system of management. The incrementalist approach, coupled with a pluralist or co-ordinated system of management, could well be satisfactory in countries that allocate abundant resources to R & D, are basically satisfied with the results of their current incrementalist policies, and consider that their current science policy machinery is equipped to deal with the pressures on resources that have arisen in recent times. If, however, these conditions are not satisfied, there is clearly a need to re-appraise systems and to continue the search for more 'rational' approaches to making decisions and setting priorities.

Analytical methods in science policy

At the beginning of this chapter it was suggested that one of the consequences of the success of the scientific enterprise was the belief that scientific thinking and methods could be applied to decisions about science policy. The subsequent discussion has indicated that this has not so far proved to be the case and that the criteria that have been developed after many years of deliberation about the problems of priorities and resource allocation are subjective, qualitative, non-analytical and incrementalist. Clearly, if a rational, objective approach to science policy is possible, then decision-makers must be able to assess the costs and benefits of particular courses of action in a quantitative way.

It has been noted that the most significant factor that shaped contemporary attitudes towards science was scientists' success in military operations during the Second World War, and that war also demonstrated the value of scientific thinking in dealing with strategical and tactical problems. The new breed of operations researchers, led by J.D. Bernal, showed that even in the 'non-scientific' arena, scientific methodology could bring order out of chaos—the boffin was able to teach the military a great deal about its own business. In the post-war era, with growing international trade and increased economic competitiveness, a strong positive attitude towards managing scientific organizations coincided with greater government involvement in the conduct and administration of scientific research. It was natural to assume that if science was generating economic progress, the economy should be managed in a scientific way.

In the 1960s government spending had reached the point where all the claims on the available resources could not be met, even in the United States, and it is there that analytical techniques were first introduced into government departments in an effort to control government spending and to choose programs and projects that deserved support and eliminate bureaucratic rivalries. In 1966, President Johnson decreed that the Programming, Planning and Budgeting System (PPBS) would be used for

• selecting or identifying an organization's overall long-range objectives and analysing systematically the various courses of action in terms of relative costs and benefits (planning);
• deciding on the specific courses of action to be followed in carrying out planning decisions (programming);
• translating planning and programming decisions into specific financial plans.

The PPBS system was the embodiment of rationalist decision-making. In the late 1960s and early 1970s, other factors emerged which, on the one hand, increased the need for formal analytical techniques to assist the decision-makers but, on the other, brought into question the value of the techniques, including PPBS, that were in use or had been used in industry and government. Economic growth rates began to decline; government spending was increasing; the public was becoming more aware of the need to protect the environment, curb pollution, conserve resources and halt the march of job-displacing technology. These factors all pointed to the need for better analytical and forecasting techniques, but the results that were emerging by then were not encouraging. The complex area of management and administration was not as amenable to analysis as the less complicated area of production processing. The Government's faith in PPBS began to diminish after the huge cost overruns that occurred with large military projects like the McDonnell-Douglas F111 fighter aircraft, where resources were committed before the technical problems had been solved. In 1970, PPBS was abandoned.

In 1969, the OECD undertook a review of forecasting and other analytical techniques that were considered to be of potential use in science policy. The subsequent report concluded that:

- important decisions rarely can or should be made solely on the basis of quantified cost–benefit analysis, simply because in nearly all cases neither costs nor benefits can be made commensurable except through technical assumptions that inevitably hide political value judgments;
- many government-financed R & D activities are intended to have a pay-off only in the longer term. This means that the results of any formal analysis will be subject to arbitrary assumptions about the time–discount rate and about what the world will or will not be like in the distant future;
- scientific and technological activities often contribute to policy objectives only in combination with other activities — for example, expenditure or physical capital. This makes it particularly difficult to estimate, either before or after the event, science and technology's true contribution to achieving a policy objective;
- it is difficult to predict the outcome and the socio-economic and political effects of scientific and technological activities; we simply do not yet know enough about the interaction between science and technology on the one hand and social, economic and political processes on the other.[39]

Despite these shortcomings, the report acknowledged that formal analytical techniques have certain advantages. They facilitate the systematic examination of alternative policies and the means of implementing them. When known facts about and predictable consequences of any policy are presented, policy-makers and analysts are able to isolate the factors requiring technical and political judgment. Frameworks can be established for periodic review and evaluation of scientific and technological progress, the possible consequences of that progress and the possible consequences of adjustments made in the light of changing knowledge and circumstances. But, as Pavitt puts it, 'while one can argue probably that the use of formal analytical techniques can lead to better decisions, one can equally argue that they can lead to worse decisions.'[40] Brooks, in discussing analytical methods in general, said this:

> It would be hard for me to say that the techniques are absolutely useless because I think it is probably a valuable conceptualization of the problem [of resource allocation]. But the actual application of mathematical analysis to decision-making has had zero impact and probably deservedly so.[41]

A full-scale review of analytical methods is beyond the scope of this book, but in the following sections the more important ones are described briefly. Many different analytical methods in use at the present time might be of use in science policy. Some are quite specialized, whereas others can be applied generally, and there are various ways of classifying them. Jansch distinguishes exploratory from normative techniques; that is, those that examine the influence of technological trends on future policy objectives and those that take the policy objectives as their starting point. The techniques can also be classified according to their application, which may be long-term or short-term forecasting or monitoring of operations. Forecasting techniques include trend extrapolation, causal models, expert consultation, the Delphi technique, scenario construction and cost-benefit analysis.

TREND EXTRAPOLATION

Trend extrapolation involves predicting future trends in the system under study by examining the past and making conditional statements and assumptions about the future. It is simple and rapid, and for short-term forecasts within well-defined limits it is relatively reliable. For longer-term forecasts or the analytical data needed to formulate policy, however, it is quite deficient, because underlying causes are ignored and continuity between the recent past and the foreseeable future is assumed. As an aid to making policy decisions, it is also less than satisfactory, because if the trends are followed, then the policy-makers cease to shape external events in the light of what is needed and allow the *status quo* to continue unchecked. Curve-fitting is used to refine the method. The underlying assumption is that there is a simple mathematical relationship between variables in the past that will hold for the period of the forecast, and if the resulting graph is extrapolated into the future, trends can be predicted.

CAUSAL MODELS

As the name suggests, the causal technique involves forecasting trends on the basis of cause. For example, a prediction about future population levels will be based not on past trends, but rather on underlying causes of population growth or decline such as fertility rates, local customs, levels of nutrition, pollution, age/sex distribution and so on.

The linear model of innovation discussed in Chapter 2 is a causal model.

The causal approach can be very helpful in evaluating alternative policies. In many cases, however, causal factors are difficult to identify; as a result, the model may be inaccurate or there may be unstated technical assumptions or shaky functional relationships that are hidden from the user. The technique relies on the future being free of surprises; unexpected events, like the oil crisis of 1973 or the tendency towards reduced technological innovation that has become endemic in Europe, cannot be accommodated.

RANKING METHODS

A number of different techniques fall under the heading of ranking methods, but they all have in common the attempt to obtain consensus on goals and the means of achieving them. Ranking methods are valuable in that they promote the exchange of information between participating experts, and this can lead to better advice and decision-making. However, the technique suffers from some inherent defects. To begin with, the relationship between the priority given to an objective and the means of achieving the objective is ignored. Also, goals and objectives can rarely be considered in isolation, especially within the political sphere. Governments have many objectives, some of which will reinforce one another while others will conflict. The objective of reducing petrol consumption, for example, may conflict with that of controlling pollution, because the devices used to control motor vehicle pollution increase petrol consumption.

EXPERT CONSULTATION

The best known expert consultation technique is Delphi, which is based on the assumption that deep in the collective consciousness of a group of people expert in a given area are the elements of an accurate forecast of developments in that area. The Delphi technique seeks to extract the forecast with a minimum of contamination from other participants or from the social environment. The experts are sent a questionnaire listing possible developments in the area and are asked when, if ever, the developments will take place. The list may be generated by the organizers of the Delphi forecast or it may be compiled from lists supplied by the respondents. The responses to the questionnaire are then summarized, the summary circulated to participants and their comments sought. Sometimes, in the light of other participants' comments or as a result of estimates the summarizers consider to be representative of group opinion, individuals change their own estimates. The question that must now be asked is — does the Delphi technique produce accurate forecasts? Very little empirical work has been carried out to assess the accuracy of such forecasts in the area of science and technology policy. Grabb and Pike[42] compared six Delphi forecasts of computer developments and concluded that, in general, these forecasts were far too conservative. In other evaluations, there were widely divergent opinions on the timing of future events.[43]

The Delphi technique has been criticized on several grounds. First, the assumptions that intuition provides a sound basis for forecasting and that interaction between the participants is under the analysts' control are unwarranted. Second, experts on the present state of knowledge in a particular area are not necessarily those who are best able to predict new developments in that area. Third, there is no guarantee that the respondents will all interpret the questions in the same way or that answers are not biased by the way the questions are presented or influences from previous questions. Finally, whereas it is claimed that Delphi avoids the dangers involved in relying on one expert's opinion and overcomes the innate conservatism of committees, it is subject to 'bandwagon' effects, and there is no proof that consensus of opinion of a group of experts is any better than the opinion of one — individuals or minorities are often right.

SCENARIO CONSTRUCTION

The widely used technique of scenario construction, developed by the RAND Corporation in the 1950s, attempts to evaluate policy options by considering in detail the implications of each option with various projected outcomes. Where there are relatively few outcomes — options of reasonable certainty — the technique can be quite reliable. It does, however, depend on experts and suffers from the limitations previously discussed in relation to experts. It also relies on the future being free of surprises — Herman Kahn, the famous Hudson Institute 'futurologist', considered many scenarios, but he failed to foresee the possibility that oil prices might double within five years of the publication of his famous book *The Year 2000*.

COST–BENEFIT ANALYSIS

The technique of cost–benefit analysis is used to decide between paths to a particular objective in the relatively short term. It attempts to measure in money terms the costs and benefits to be expected from projects over a period of time, setting out the factors that need to be taken into account in making certain economic choices.[44] The important first step in carrying out a cost–benefit analysis is to decide whether the technique is applicable, since choices may be limited at the outset by political or other factors. Other steps include setting a time limit and preparing a financial estimate. Time limits tend to be arbitrary, and since projection into the future is inevitable, the further this projection is, the more likely it is that errors will be introduced. The

financial estimate, too, may be quite inaccurate, especially in the case
of highly technological projects where the technology is not fully
developed. The huge cost overrun in the Concorde project is a case
in point. Further steps in a cost–benefit analysis include distinguishing
costs from benefits, quantifying and weighting costs and benefits,
discounting and assessment.

Few of these steps are free from serious problems. In a major review
of forecasting techniques, a group at the Science Policy Research Unit
at the University of Sussex concluded that 'it may be that the only time
cost–benefit studies should not be undertaken is when the results may
be taken seriously. It is all too easy to forget the defects and
arbitrariness of the method.'[45]

OPERATIONAL TECHNIQUES

Operational analytical techniques are aimed at implementing
programs efficiently. Typically, such techniques are geared to prevent
cost and time overruns and to ensure that departures from planned
performance are brought to the program administrators' attention.
The most famous of these techniques are Program Planning and
Budgeting System (PPBS) and Program Evaluation and Review Tool
(PERT). PPBS was introduced into all United States government
departments by President Johnson in 1966. It was an ambitious system,
which required objectives to be specified and alternative programs to
be costed and evaluated in the context of a multiyear plan linked to the
budget cycle. As has already been indicated, PPBS was not a success
and was phased out in 1970. PERT, developed by the RAND
Corporation, has been more successful than PPBS, though its critics
point to the cost and time overruns in the Polaris development and to
its use at Lockheed in that company's Tristar development to show that
it, too, has severe drawbacks.[46]

In the following chapters, elements of the R & D systems of a
number of OECD countries will be described. While an attempt will be
made to classify the science policy machineries of these countries in
terms of the four systems described in this chapter, it will be seen that
no country matches any of these systems exactly. The countries are
loosely arranged in order of the increasing relative co-ordination of
their R & D activities, the United States having the least and Belgium
the most. The choice of countries was dictated by the initial brief of the
Australian Science and Technology Council described in the preface.
The advisory, executive and operational institutions that have had the
greatest impact in shaping the R & D systems of these countries will be
described and the way in which R & D priorities are set will be
discussed where appropriate.

CHAPTER 4

Science policy in the United States

Separation of powers is a characteristic of the American political system, with a system of checks and balances that ensures the coherent working of the three independent branches of the government — the executive, the legislature and the judiciary. This separation of powers symbolizes American pluralism, a belief that no special interest should command a majority in the nation and that there should be a plurality of competing interest groups and diversity of rival interests. In keeping with this ethos, the government's R & D system is not controlled by a single monolithic operational department, nor is there a central Science Policy Department to program and plan the nation's research effort or develop a coherent science policy, though there is an Office of Science and Technology Policy within the Executive Office of the President. Many departments and agencies are responsible for maintaining the government's R & D effort and these work independently of one another. General policy is formulated by the President in consultation with his advisers within the Executive Office of the President (see Figure 4.1), but once the departments and agencies have received their appropriations through the budget process it is up to them to decide on internal distributions. This is the classic pluralist science policy system; the nation's science policy is the *ex post facto* sum of the science policies of the various elements of the R & D system. The government's role is to stimulate research, not to direct or control it in any rigid way. Individual departments are apt to attack a common problem from different points of view, using different tactics, and this pluralist approach, which the United States can afford, is widely believed to be the source of the nation's supremacy in scientific research and technological development across a wide variety of fields and disciplines.

It would be misleading to say that the vast American scientific enterprise is completely unplanned and uncoordinated or that the government has not attempted to institute planning and co-ordinating mechanisms. It was pointed out in Chapter 1 that when the National Science Foundation was established in 1950 the enabling legislation

Figure 4.1

Main elements of the U.S. Federal Government's R & D system

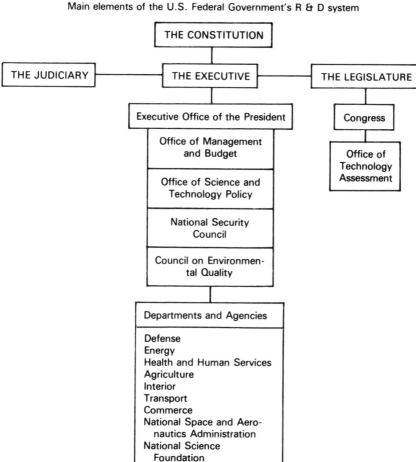

contained clauses that gave the Foundation a clear policy role. Apart from its funding responsibility it was supposed to develop and encourage the pursuit of national policy for the promotion of basic research, determine the impact of research upon industrial development and the general welfare and evaluate the scientific research programs undertaken by agencies of the Federal Government. As Albert H. Teich, science policy analyst at the American Association for the Advancement of Science points out, though the word 'planning' does not appear in this mandate, the functions had a distinct planning flavour.[1] For reasons that have already been explored, the National Science Foundation did not adopt the role of science planner even though the

Bureau of the Budget was anxious that it should do so. Even President Eisenhower's Executive Order of 1954, formulated in order to reinforce the Foundation's policy, evaluation, and planning roles, failed to do this, for it was not given the necessary authority to force the big departments and agencies to co-operate in the evaluation exercises it was supposed to carry out; nor was there any precedent for an arrangement whereby a small agency would exert authority over much stronger departments, although they were theoretically on an equal footing. It was only by establishing a supra-departmental agency or office within the Executive Office of the President that such a mandate could be carried out and, as noted in Chapter 1, such an institution was created when the Science Advisory Committee was reconstituted in the wake of the Sputnik triumph in 1957 as the President's Science Advisory Committee and moved to the White House. Two years later a Federal Council for Science and Technology, composed of government officials, was set up to co-ordinate the R & D activities of the departments and agencies. Both of these advisory bodies were chaired by the Special Assistant to the President on Science and Technology, a post that was upgraded to Presidential Science Advisor in 1962, when an Office of Science and Technology was created within the Executive Office of the President. However, because of its purely advisory status this science policy apparatus was relatively weak and ineffective. In 1973 it was abolished by President Nixon and the Director of the National Science Foundation became the President's Science Advisor.

In 1976, responding to the protests of the scientific community and of the National Science Foundation itself, the Ford Administration re-established a scientific presence in the White House by creating the Office of Science and Technology Policy (OSTP). In 1982 a Science Council, reporting to the Director of OSTP was established in order to improve co-ordination of the national research effort.

The Executive Branch

Research and development in the United States is performed mainly in the government departments and independent agencies, in the universities and in industry. In Table 4.1, the estimated total national pattern of R & D for 1982 by performer and source of funds is shown. It can be seen that whereas the Federal Government is the source of nearly half of all R & D funds in the United States, it spends a mere 13 per cent of the gross national expenditure of $77 billion (thousand million) in its own laboratories. Industry was expected to receive almost $18 billion from the government departments and agencies and to contribute $38 billion of its own to R & D, making this the largest performance sector. It is not without reason, therefore, that the United

Table 4.1

Distribution of US GNERD for 1982 (estimated)

Source of funds / Performer	Federal Gov-ernment	Industry	Colleges and universities	Non-profit institutions	Total
	$ US million				
Federal Gov-ernment	10 000	—	—	—	10 000
Industry	17 800	37 900			55 700
Colleges and universities	4 600	275	1 600	475	6 950
Federally funded R & D centres	2 350	—	—	—	2 350
Non-profit institutions	1 375	325	—	585	2 285
TOTAL	36 125	38 500	1 600	1 060	77 285

Source: R & D Report VII Shapley *et al.*, p. 12

States has been called the 'contract state' since contracting to industry and to the universities accounts for most of the allocations to R & D.

It is necessary to know the final steps in the budgetary process in order to understand how the allocations are made to the departments and agencies. First, the Office of Management and Budget prepares a target budget, taking into account the economic situation and the multiyear projections of the previous budget. Then the departments and agencies make their budget requests under various headings, including R & D, and the Office of Management and Budget assesses these requests. After hearing the appeals against the decisions it has made about them by the departmental and agency heads, it makes its recommendations to the President. Research and development proposals are submitted by the departments and agencies with all their other requests, but in considering them, the Office of Management and Budget seeks the advice of the OSTP. A unified budget is presented to Congress by the President in January and this is reviewed by authorizing and appropriations committees of each House of the Legislature. The authorizing committees define the program boundaries and funding limits; the appropriations committees can only recommend funding for those programs that have been approved by the authorizing committees and cannot exceed the financial ceilings imposed. The authorizing and appropriations committees delegate the budget review to subcommittees, each of which is responsible for one

or more agencies. During the period of final consideration departmental and agency representatives are called before the subcommittees
to explain or justify their budget items and after the vote in Congress
the budget authorities are issued.

When President Carter's fiscal year (FY) 1980 budget was presented
to Congress in January, he followed it up with a Message to Congress on
Science and Technology,[2] an important initiative which has not been
pursued by President Reagan. In this Message, which was meant to
assist the Congress in its task of considering the R & D programs of the
departments and agencies, Carter

- described the Administration's policy perspective on science and
  technology and the roles of government, industry, universities, and
  the public in support of science and technology;
- highlighted some of the most important science and technology
  initiatives undertaken in the Administration in domestic, national
  security and international arenas;
- outlined the potential effects of science and technology on the
  nation for the 1980s.

The importance of the budget process in establishing priorities was
underlined by William Carey, Executive Officer of the American
Association for the Advancement of Science when he said:

> The Federal Government holds most of the high cards which determine the
> thrust and priorities of scientific research and development effort in the
> United States. Its policy goals and budget preferences to a very great extent
> have shaped the environment in which science and technology have been
> carried on since the 1940s. With the possible exception of basic science
> 'research and development' *per se* are not straightforward objectives of
> government policy but are means towards the support of other primary
> objectives. The budget outcomes, accordingly, are for the most part deri-
> vative of a highly complex and somewhat arcane decision-making and
> bargaining process which is embedded in a maze of institutions involving
> both the Presidential and the legislative systems.[3]

An indication of the Administration's priorities can be obtained from
an analysis of the budget documents and of the President's Message to
Congress. The OSTP plays an important role in the exercise that sets
these priorities.

The Office of Science and Technology Policy

The Office of Science and Technology Policy (OSTP) was established
by law in 1976 to 'serve as a source of scientific and technological
analysis and judgment for the President with respect to major policies,

plans and programs of the Federal Government.'⁴ The Director is the President's chief scientific adviser and his duties are:

- to provide within the Executive Office of the President, advice on the scientific, engineering and technological aspects of issues that require attention at the highest levels of government;
- to advise the President of scientific and technological considerations involved in areas of national concern, including but not limited to the economy, the environment and the technological recovery and use of resources;
- to evaluate the scale, quality and effectiveness of the federal effort in science and technology and to advise on appropriate actions;
- to advise the President on scientific and technological considerations in the federal budget, assist the Office of Management and Budget with an annual review and analysis of funding proposed for R & D in the budgets of all federal agencies and aid the Office of Management and Budget throughout the budget development process;
- assist the President in providing general leadership and co-ordination of the R & D programs of the Federal Government.

The concept of an Office of Science and Technology Policy as set forth in the Act collided head-on with President Carter's views on how the Executive Office of the President should function. He was committed to a substantial reduction in Executive Office staff and the wide array of functions specified for the OSTP would necessitate a large increase. The President and his close advisers wanted to avoid giving special interests a foothold in the White House and OSTP was regarded by many as an organ of special pleading for science in the Executive Office. After apparently giving serious consideration to outright abolition of OSTP when he assumed the Presidency, Carter decided to retain a small OSTP and to abolish or redefine some of the more elaborate and cumbersome functions and machinery specified in the Act, taking into account the fact that Congressional approval would be needed to abolish it and a confrontation with the Congress in the early days of his Administration would be unwise. Two important planning functions became immediate victims—the Five Year Outlook and the Annual Report on Science and Technology. According to the Act, the Annual Report was to

- review developments of national significance in science and technology;
- discuss the significant effects of current and projected trends in science and technology on the social, economic and other requirements of the nation;
- review and appraise selected science— and technology—related

programs, policies and activities of the Federal Government;

- compile an inventory and forecast of critical and emerging national problems the resolution of which might be substantially assisted by the application of science and technology;
- identify and assess scientific and technological measures that could contribute to the resolution of such problems in the light of the related social, economic, political and institutional considerations;
- review the existing and projected scientific and technological resources, including specialized manpower that could contribute to the resolution of such problems;
- recommend legislation on science— and technology—related problems that would contribute to the resolution of such problems.

The Five Year Outlook was to identify and describe situations and conditions which warranted special attention within the following five years involving:

- current and emerging problems of national significance that were identified through scientific research or in which scientific and technological considerations were of major significance;
- opportunities for, and constraints on, the use of new and existing scientific and technological capabilities which could make a significant contribution to the resolution of the problems already identified or to the achievement of federal program objectives or national goals.

Responsibility for the Annual Report and the Five Year Outlook was transferred by the new President to the National Science Foundation (NSF) thereby destroying their credibility as major planning documents for, as Teich has pointed out, in order to carry any weight such documents need the institutional support of the Presidency.[5] The NSF has taken responsibility for compiling the Annual Report itself but the first Five Year Outlook was contracted to the National Academy of Sciences.

The OSTP has remained a relatively small unit within the Executive Office, dealing with the budget and other short-range matters, where its influence is very great. As a long-range strategic planning unit, however, its influence is negligible. Its view on priority identification is that, given the way in which the American R & D system has evolved, there is no possibility that the Office can become involved in any overall priority-identification exercise. It does attempt to review specific areas, however, and make recommendations in the budget context. No attempt has ever been made to develop or use any of the analytical techniques discussed in Chapter 3 for priority identification. The Office uses a task force approach in its reviews in which working groups with broadly-based memberships examine specific issues; a working group on basic research in the Department of Defense was drawn from the universities, industry and government departments, for example. Long-term forecasting or priority setting are not favoured because it is widely held within the organization that forward plans must be coupled to the political process. Projections too far ahead are made in a political vacuum and can have little impact. This is not a view that is shared by all, however. Teich has this to say:

> It is likely that there will always be tension in the American R & D system, as in other areas of American society, between the desire to maintain pluralism and the demands for more comprehensive control, planning and management. Few United States policy makers would like to risk losing the many benefits that derive from the diversity of the existing system. Nevertheless resources for R & D are increasingly constrained, and it is essential that the United States find more effective and efficient means for their allocation. It does not seem impossible to maintain the virtues of pluralism while improving the nation's capability for strategic planning in science and technology.[6]

Nevertheless, the development of a long-range strategy has been removed to the National Science Foundation and, through it, to the National Academy of Sciences.

The National Academy of Sciences

The National Academy of Sciences is a private, honorary society of scientists and engineers dedicated to the furtherance of science and its use for the general welfare. It is not a federal agency although it is called upon by the terms of its charter to examine and report on any subject of science and technology upon request of any department of the Federal Government. It occupies, therefore, a unique position among scientific societies because it operates under a century-old federal charter to provide scientific and technical advice to the government, while remaining independent of governmental control. Its independence enables it to provide unbiased expert advice to the various federal agencies without risk of political pressure. Whether contracted by the Federal Government or supported by private organizations, analytical studies are conducted by committees established in the Institute of Medicine and the National Research Council. The latter is the 'operating arm' of the Academy of Sciences and the Academy of Engineering. It was established under the National Academy of Sciences charter in 1916 and is now organized into multidisciplinary commissions and single-discipline assemblies. It can call on experts in any area of science and technology to serve on its numerous boards, committees, panels and working groups. The Academy is an important source of advice to the Federal Government on science and technology and its influence should not be underestimated.

The National Research Council, which issues about one report every working day, is a principal contributor to the stream of reports directed at the policy makers. A report may take the form of the proceedings of a conference, a letter to a sponsoring agency or a formal report by a project committee, and these reports are rarely ignored, as an analysis carried out within the National Research Council in 1977-78 showed. The President of the National Academy of Sciences, in presenting the analysis in the 1978 Annual Report of the National Research Council, said:

> The hopes that we had as we entered upon this analysis have been largely borne out. Well-reasoned proposals for programs that would advance the frontiers of science do receive sympathetic attention and are generally accepted. Specific technical recommendations deriving from analyses of relatively narrow technical subjects and requested by a concerned mission agency almost invariably find acceptance. Recommendations of broader character, offered in a report that has been explicitly requested by the Congress or some agency of the Executive Branch are frequently adopted.[7]

The National Academy of Sciences has had a continuing interest in the question of priority identification, especially in basic research, since 1963 when the Committee on Science and Astronautics of the House of Representatives concluded a formal agreement with it to produce a comprehensive study 'designed to throw into bold relief some of the more serious phases of policy which the government must consider in its decisions to support or otherwise foster research in America'. The House committee set out two very broad questions of fundamental importance to the Federal Government in connection with its scientific R & D program. These were:

- what level of federal support is needed to maintain for the United States a position of leadership through basic research in the advancement of science and technology and their economic, cultural and military applications?

- what judgment can be reached on the balance of support now being given by the Federal Government to various fields of scientific endeavour and on adjustments that should be considered either within existing levels of overall support or under conditions of increased or decreased overall support?

The resulting report was an important contribution to the literature on priorities. The positions adopted by the principal theoreticians are set out in Chapter 2. However, with regard to the specific questions posed by the House committee, all that could be offered was:

- more accurate statistics should be compiled on American R & D;
- additional research should be carried out on estimating the cost effectiveness ratios of research;
- on the whole, science in the United States enjoys pre-eminence and what is done in future should be based on expanding and improving the present situation;
- support for the physical sciences should be increased by one or other of two mechanisms. Either the mission agencies should devote a larger fraction of their budgets to basic science and, therefore, incline towards a broader interpretation of what kinds of basic research they deem relevant to their mission, or the National Science Foundation should become a much bigger agency, providing support for basic research that is too remote to merit support from the mission-oriented agencies.[8]

No specific criteria were developed in this exercise nor has there been any attempt since then to go beyond the 'expert opinion' method of identifying priorities. Subsequent reports of the National Academy of Sciences that delve into the problem of priorities have adopted modified Weinberg criteria; one such study is mentioned in Chapter 3.

Priorities of the Federal Government

We have said that the Federal Government's priorities can be identified by looking at the trends in the requests made to Congress for R & D in the departments and agencies and the subsequent appropriations. Research and development was considered by the Carter Administration to be so important politically and economically that a special Message on Science and Technology was presented to Congress in March 1979, two years into the life of the Administration, when its science and technology policies had begun to firm up. In the Message, the President signalled the intention of his Administration to continue the favourable treatment that basic research had received since 1977. Criticising the treatment that this area had received at the hands of previous Administrations, the President indicated that there would be real growth in expenditure, not only in the National Science Foundation, but also in the mission-oriented agencies such as the Departments of Defense, Agriculture and Energy, whose commitment to basic research seemed to have eased. Strengthening this commitment was to be a central element of the Administration's science policy. Pointing out that since his government had taken office requests for basic research had increased by 26 per cent, and that for the first time the budget request for the National Science Foundation had exceeded $1 billion, the President asked for a further increase of 11 per cent for FY 1981.

Other elements of the Administration's policy given in the Message were:

- stimulation of innovation in industry to sustain economic growth and improved productivity;
- meeting of the nation's energy, resource and food needs;
- promotion of better health for all;
- improvement of regulatory processes;
- expansion of the beneficial uses of space;
- understanding of the forces of nature, natural disasters and changes induced by man.

Most of these objectives are unexceptional but the first, the stimulation of innovation in industry, marked a significant change in the traditional attitude of governments in the United States towards the support of private industry by measures other than the contract mechanism. In October 1979, some six months after his Message on Science and Technology had been delivered, the President revealed the initiatives he had in mind in order to fulfil this objective. They arose out of a domestic policy review on innovation carried out by the Department of Commerce in 1978 and early 1979 and included the following:

- enhancement of transfer of information by establishing a Centre for the Utilization of Federal Technology;
- increase of technical knowledge by the establishment of four 'generic technology' centres by the Department of Commerce and the National Science Foundation, and by increased support for joint industry-university research programs;
- encouragement of the development of small innovative firms by enhancement of the National Science Foundation's Small Business Innovation Research Program, and by the establishment of corporations for Innovation Development by which entrepreneurs might gain access to investment capital with matching loans from the Federal Government;
- improvement of the regulatory system by the inauguration of five year forecasts of 'priorities and concerns' by the federal health, safety and environmental regulatory agencies to give industry time to develop 'compliance technology';
- encouragement of labour and management adjustment to technical change by the establishment of a Labor-Technology Forecasting

System to develop advance warning of industrial changes and permit timely adjustment;

- maintenance of a supportive federal climate through the establishment of an annual award for technological innovation and by Department of Commerce-National Science Foundation sponsorship of a national conference for deans of business and engineering schools to stimulate improved curriculum development in technology management and entrepreneurship.

The FY 1981 budget was presented to Congress by the Carter Administration at the end of 1980 after the normal round of negotiations between the Office of Management and Budget, OSTP, the agencies and the departments. When it was being put together at the end of 1979 the economic and international situations were relatively stable. Though inflation was on the increase and a recession seemed inevitable, the general mood of the Administration was relatively optimistic. But as the year drew to a close the international situation became very tense with the invasion of Afghanistan by the Soviet Union and the seizure of the American hostages in Iran. The domestic economic situation, too, showed signs of rapid deterioration. As a result of this, and of the need to make an impression on the American public in an election year, the Administration took the relatively unusual step of revising the January budget in order to bring it into balance. A new budget for FY 1981 was presented in March 1980 which sliced about $1 billion from the R & D appropriations. The 'real growth' in basic research expenditure was halted and the $50 million industrial initiatives program was cut by $14 million. But, in the FY 1982 lame duck budget, so called because it was presented to Congress by the outgoing Carter Administration and repudiated by Mr Reagan before he took office, there was a reaffirmation of Carter policies towards R & D and substantially increased requests went to Congress.[9]

The Reagan Administration which took office in January 1981 was, in the first few months of that year, preoccupied with its plans for cutting government expenditure substantially and at the same time reducing taxes, in line with the economic theories that it had adopted. The American Association for the Advancement of Science (AAAS) summarized the main policy changes reflected in the 1981 and 1982 budgets as follows:

- non-defence programs should be cut and new initiatives deferred;
- defence programs should be increased;
- the anti-nuclear bias of the Carter Administration should be halted and its energy program dismantled; federal funding for commercial

demonstrations and technology assistance should be cut back or eliminated;

- tax reductions, not federal programs, should be relied on to provide incentives for technological innovation in industry.[10]

The revisions made by the Reagan Administration to the FY 1981 and FY 1982 budget requests, which are shown in Table 4.2, provide the information on which the AAAS based its judgments on the implicit priorities for R & D. Thus, the Department of Defense R & D budget which, in recent years, accounted for less than half of the total federal funding in this area, now accounts for about 57 per cent. The Carter Administration in the revised 1981 budget had asked for $14 billion, an increase of 9 per cent in the department's appropriation; the Reagan Administration added a further $600 million in that year and in FY 1982, $1500 million were added to the Carter request.

Table 4.2

Revisions to the R & D budgets of selected agencies
for FY 1981 and FY 1982

Agency	1981 $ million			1982 $ million		
	Carter	Reagan	% change	Carter	Reagan	% change
Defence	16 725	17 322	+3.6	20 581	22 059	+ 7.2
NASA	5 537	5 523	−0.3	6 726	6 122	− 9.0
Energy	6 245	5 867	−6.0	7 080	5 524	−22.0
HHS	3 994	3 971	−0.6	4 273	4 173	− 2.3
NSF	1 008	934	−7.3	1 254	1 020	−18.7
Agriculture	816	810	−0.7	902	889	− 1.4
Commerce	387	355	−8.3	421	316	−24.8

Source: W.H. Shapley, A.H. Teich, G.J. Breslow and C.V. Kidd, Research and Development AAAS Report VI (Washington, 1981)

In the Department of Energy, non-nuclear energy programs were dramatically reduced. In his campaign, President Reagan promised to abolish the Department of Energy. So far this has not been done, but a major realignment of the department's activities has been imposed upon it by the different approach that the new Administration has towards the solution of the nation's energy problems. The respective administrations' budget figures for nuclear and non-nuclear research tell the story and these are shown in Table 4.3. The figures show a

Table 4.3

Revisions to nuclear and non-nuclear energy
R & D in the Department of Energy Budget
Authority in FY 1981 and FY 1982

R & D	1981 $ million		1982 $ million	
	Carter	Reagan	Carter	Reagan
Nuclear	1751	1825	1769	2072
Non-nuclear	2221	1707	2518	791
Total	3972	3532	4287	2863

Source: W.H. Shapley, A.H. Teich, G.J. Breslow and C.V. Kidd, Research and Development AAAS Report VI (Washington, 1981)

significant shift of resources away from non-nuclear to nuclear energy research, reversing the trend of the Carter Administration. The Reagan policy is interpreted by the AAAS as indicating, first, a greater reliance on market forces to govern the use of energy and to dictate choices in the use of energy technologies, and second, concentration on high-risk, long-term technologies such as the fast breeder and fusion reactors, areas where the private sector lacks sufficient incentive to invest.[11]

The activities of the National Science Foundation that have been most seriously affected are the programs for the stimulation of industrial innovation which were inaugurated during the term of office of the Carter Administration. Because of the importance of these initiatives to the general theme of this book, the role and function of the National Science Foundation and its attempts to respond to the priorities coming from Capitol Hill will be reviewed briefly. The National Institutes of Health, too, has been conscious of the need to respond to the signals that have emerged from the Executive Office and the legislature.

The National Science Foundation (NSF)

The origins of the NSF have been discussed in Chapter 1. Its purposes are

- to increase the nation's base of scientific knowledge and strengthen its ability to conduct scientific research;
- to encourage research in areas that can lead to improvements in eco-

nomic growth, energy supply and use, productivity and environmental quality;
- to promote international co-operation in science.

The Foundation consists of a National Science Board of twenty-four
members and a Director, each appointed by the President. It carries
out its mandate by supporting

- fundamental and applied research in all scientific disciplines
  through grants, contracts and other agreements awarded to
  universities, non-profit and other R & D organizations;
- co-operative efforts by universities, industries and government and
  promotion of the application of R & D for better products and
  services which improve the quality of life, create employment
  opportunities, stimulate economic growth and increase productivity
  and foreign trade;
- national centres where large facilities (in astronomy and high energy
  physics) are made available to qualified scientists;
- development and use of computer and other methodologies and
  technologies;
- experimental efforts to improve and accelerate the application of
  R & D by industries, government and other institutions;
- measures to bring about a greater understanding of science and
  technology as it affects contemporary life, including the social and
  ethical implications of an increasingly technology-dependent
  society;
- research aimed at formulating science policy through analysis of
  existing and emerging national issues that have significant scientific
  and technological content.

The Carter Administration's FY 1982 budget requested an increase of
more than 20 per cent in the NSF's appropriation, from $1.03 billion
in FY 1981 to $1.26 billion in 1982. However, in keeping with its
policy of reducing government expenditure other than on defence, the
Reagan Administration reduced the 1982 figure to $1.02 billion, a cut
of nearly 19 per cent. However, this cut did not fall on the basic
research element of the NSF's budget which, at $950 million, was 11
per cent up on the 1981 figure. This trend is reflected in the Office of
Management and Budget's statement that 'research in the natural
sciences and engineering is of relatively high importance to future
technological advancement and the long-term economic health of the
nation.'[12]

   In contrast, the social and behavioural science programs of NSF
were severely pruned. The Carter budget request of $40 million for
social and economic science in FY 1982 was reduced to $10 million,

and the $43 million request for behavioral and neural sciences was reduced to $29 million. The AAAS has interpreted the reductions in social, economic and behavioural sciences as due to ideological bias rather than a comment on the quality of the research itself.

PRIORITIES OF THE NSF

The NSF is organized into a number of directorates, each being the responsibility of an Assistant Director. Within the directorates there are varying number of divisions. For example, in the Mathematical and Physical Sciences Directorate there are four Divisions— Mathematical Sciences, Computer Research, Physics and Chemistry and Materials Research. Allocation of resources to the directorates is not carried out by any formal priority-setting technique. Some twelve months before the beginning of a fiscal year, the NSF Director is asked by the Office of Management and Budget to prepare a preliminary budget, and he is given a target figure. The Director, in turn, asks his Assistant Directors to prepare their budgets, and these are debated in the presence of the Director until consensus of a sort is reached. In justifying their budgets the Assistant Directors use all means at their disposal, but these rarely, if ever, involve analytical techniques of the sort discussed in Chapter 3. The traditional model of advocacy is used, which relies heavily on inputs from the scientific community. At one level, this takes the form of 'proposal pressure', the stream of incoming unsolicited research proposals that provides an indication of the priorities, research interests and problem orientation of the practitioners in a particular field. These constitute a measure of demand from the Foundation's scientific constituency. At another level, studies and reports by the Foundation's advisory committees or groups within the National Academy of Sciences give more general guidance on the directions that the NSF might follow in its granting function. The information produced by these processes as well as by discussions of the National Science Board is assimilated by the programming staff and out of these interactions emerge a set of implicit priorities, program thrusts and specific decisions on proposals. The following criteria are said to be used in selecting projects:

- Category A: criteria relating to the competent performance of the research—the technical adequacy of the performer and of his institutional base;
- Category B: criteria relating to the internal structure of science itself;
- Category C: criteria relating to utility or relevance;
- Category D: criteria relating to future and long-term scientific potential of the United States.

Because none of these criteria are susceptible to precise quantification or even in most cases to unambiguous rank ordering, it would be more accurate to speak of 'factors considered' rather than of 'criteria'. In addition, different relative weights must be attached to the factors in the case of different agency objectives or programs. Criteria in Category A are applied in every proposal. Every NSF-supported project is expected at the very least to produce some valid new information or relationships. Peer evaluation is clearly the technique used in this and the following category. The criteria of relevance and utility are threefold:

- the probability that the research can serve as the basis for new inventions or improved technology;
- probable contribution of the research to technology assessment;
- identification of an immediate programmatic context and user of the unanticipated research results.

In the NSF's opinion, the criteria of relevance cannot be made entirely distinct from the preceding two since

> that science judged best by internal standards has almost invariably turned out in the long run to be the most useful. Valid generalizations and powerful methods of observation and measurement usually lead to new invention, improved technology and more confident assessment. Conversely, applied investigations designed to support invention technology and assessment tend to succeed in these purposes to the extent that they do uncover valid generalizations or improved methods.[13]

Criteria relating to long-term potential are seldom dominant in project selection of program development but they are always considered. The policy of the Foundation is not to undertake any action for short-term reasons that would seriously jeopardize the long-range scientific potential of the nation. Criteria under Category D are listed as

- probable influence of the research upon the capabilities, interests and careers of junior participants;
- probability that the research will lead to radiation and diffusion, not only of technical results but also of standards of workmanship and a tradition of excellence in the field;
- anticipated effect upon the institutional structure of American science.

Thus, it is claimed that within the NSF scientific merit is not the only criterion that is used in the evaluation of proposals and referees are asked to evaluate them by taking into account the various criteria set out above. One suspects that in view of the convenient 'collapse' of criteria of relevance and utility into the criterion of scientific merit, it is the latter that is dominant in the decision-making process, at least in the non-applied directorates.

APPLIED RESEARCH IN THE NSF

In 1968, the NSF Act was amended to allow it to fund applied as well as basic research and, in 1969, the Foundation established a program for Interdisciplinary Research Relevant to Problems of Our Society (IRRPOS). This was in response to a general feeling within the legislature that the enormous amout of basic research that had been funded in the 1960s by NSF and the other departments and agencies was not being applied to the growing social and environmental problems that were then beginning to attract considerable attention in the nation. IRRPOS was on rather a small scale; $6 million were allocated in 1970, and $13 million in 1971. In 1971, however, the Office of Management and Budget proposed to NSF that $50 million would be available for an aggressively managed active program of applied research. The Research Applied to National Needs (RANN) Program was the outcome of this proposal, and Dr Alfred Eggers, a

Table 4.4

Criteria for judging the suitability of
proposals to RANN

Importance	Where the significance and urgency of the problem area or the potential consequences for the nation were great.
Pay-off	Where domestic, economic and social benefits to be realized were significantly higher than the anticipated research and implementation costs, or where the potential to strengthen the United States' international position existed.
Leverage	Where science and technology could have unique and substantial impact on the problem.
Readiness	Where the effort was timely and scientifically ready and the skilled manpower was available.
Capability	Where federal, academic and industrial capabilities existed to mount a successful research program.
Need for federal action	Where the research is not being conducted by private industry because the immediate incentive is not sufficient or the market is fragmented.
Unique position of NSF	Where the NSF could most effectively serve the research needs of the government.

former National Aeronautics and Space Administration (NASA) official, became its first Assistant Director. A set of criteria for judging the suitability of research projects was drawn up, and these are shown in Table 4.4, together with the interpretations that were put on them. The major program elements identified in the initial stages of RANN were (i) environmental systems and resources, (ii) social systems and human resources and (iii) advanced technology applications. The criteria and program elements were determined by a task force, which perused various agency reports on the use of science for the achievement of national objectives and related the problems that were contained therein to ongoing NSF activities that could be classified as problem-oriented.

The annual RANN appropriation increased steadily from 1972 onwards. By 1975 it stood at a mere $143 million but by this time the program was attracting a great deal of criticism. The beginning of the end occurred in that year when $51 million were removed from its budget appropriation and transferred to a new agency, the Energy Research and Development Agency, which eventually became the Department of Energy. In 1976 the budget was reduced to $60 million and in 1978 RANN as a program description within NSF was abolished. It was replaced by a new, smaller, Applied Science and Research Applications Directorate. Some activities of RANN were assigned to the new directorate and some were transferred to other NSF directorates or to other agencies.

It is beyond the scope of this book to detail the many reasons for RANN's failure. It is an important episode in the history of American science policy, since it was an attempt by an agency that traditionally had no 'mission' other than the support of high quality scientific research to extend its operations into the applied science arena. In its 20 years of operation the NSF had developed a style of management and administration that relied on inputs from its customers—the grantees in the scientific community—for pointers to areas that needed support. In the case of RANN, though the customers still came from the universities, and had the values and concerns appropriate to a research-oriented community, they were not the end-users of the research and their guidance on priorities could not be relied upon. As a senior policy analyst at NSF put it:

> (RANN) was of a free standing character since, unlike the traditional 'mission-oriented' agency, it had no set of clearly identified end-users or decision-makers whose functional concerns could provide a ready made basis for a research agenda.[14]

Thus, in most applied research programs the mission of the agency provides guidelines for the research agenda and criteria for establishing priorities. These guidelines arise out of the functional concerns of the agency, its program needs and the needs of its clients. Most of these agencies have their own laboratories and scientists who both produce research and help evaluate and guide work that is carried out extramurally. The NSF has no laboratories and its main mission was not coincident with that of RANN. The traditional methods of proposal pressure and peer review which were used to evaluate proposals and indicate the areas that needed support in the RANN program were inappropriate in an applied context. The customers for the research were supposed to call the tune, not the producers, and it was by no means certain that the producers in the universities, organized as they were into departments and disciplines, could respond effectively to a call for research applied to national needs.

In Schmandt's review of science policy research, referred to in Chapter 3, the need for institutional differentiation in the performance of research tasks was identified as a policy constant, and the consequent need to examine policies, regulations and funding practices for their impact on the different institutions performing the research. The imposition of research tasks or their encouragement in institutions that are organizationally or structurally incapable of performing them is contrary to this principle and RANN can be taken as an example. Another example involving the universities is NASA's Sustaining Universities Program (SUP). The Sustaining Universities Program was a NASA attempt to relate university research to the needs of society, seeking to link portions of the scientific community into a research system whose output would be information and technology of immediate relevance to problems of national concern. At the time of SUP's inception, NASA had a close relationship with the universities, to which it contracted a great deal of work on aeronautics and space science. In 1962 the agency decided to encourage the universities to undertake research applied to broader, more 'national', needs rather than those specific to the space program. Funds were set aside for NASA fellowships, research grants and buildings, and the terms and conditions under which the universities would receive support under the scheme were laid out in *Memoranda of Understanding*, signed by the participating university administrators. One of the more important clauses in the *Memoranda* was the requirement that the universities should seek ways of diffusing the results of space-related research by putting the funds to work in advancing development and solving technical, economic and social problems in the universities' own regions and in the nation as a whole.

By certain standards the SUP was a success; about $200 million were allocated to the participating universities between 1962 and 1970. However, the multidisciplinary aspect of the SUP was not taken seriously by the participants: they perceived the grants as institutional support in the conventional sense, not requiring innovations in the administration of research. The *Memoranda* were ignored and the program goals set out within them were not achieved because the universities failed to respond to the demand for multidisciplinarity and technology transfer. The SUP was terminated in 1970 and in the post-mortem the blame for its failure was laid squarely at the doors of the universities. If blame could be attached to NASA it was simply in terms of its naivety in expecting the universities to change. As Blankenship has put it:

> The NASA director undoubtedly believed that anything was possible. He was able to lead America to the Moon, but he could not change the university.[15]

The leverage exercised by NASA — money, buildings and agreements — was insufficient to effect changes in the universities' operations. This was an attempt to change the universities' traditional habits from the outside, to influence their priorities. But it was left to the people inside the system to choose the manner in which the 'reform' was to proceed. The SUP's clients were in an environment that worked against the aims of the Program.

Though the RANN program itself was not a success, it gave rise to a number of worthwhile developments within the NSF which put it in a good position to respond to President Carter's initiatives in the promotion of industrial innovation areas. When RANN was abolished the terms of reference of the new Directorate created in its place were to

- focus more sharply the Foundation's activities on a selected number of research applications where it could make a unique contribution;
- strengthen the links between basic and applied research activities of the NSF;
- improve the links between research applications activities and user groups in state and local government, industry and the other Federal Government departments and agencies;
- provide a funding source for high quality applied research proposals from the scientific community where such research fell outside the scope of the Foundation's basic research-funding Directorates, the focused research programs of the Applied Science and Research Applications Directorate or the research programs of the other agencies.

Five Divisions were created to carry out the Directorate's work — Problem Analysis, Integrated Basic Research, Applied Research, Problem Focused Research Applications and Intergovernmental Science and R & D Incentives. The Division of Integrated Basic Research funded topics which had the following characteristics:

- high relevance to existing or emerging problems;

- high probability that additional basic understanding would contribute to long-range solutions of the problems;
- long-term application potential;
- existence of qualified manpower.

The Applied Research Division sought to accelerate the rate of technological innovation growing out of significant advances in selected fields of science and engineering, and to provide information of relevance to public policy issues requiring a high degree of scientific input.

The Problem Focused Research Applications Division (PFRA), which received $25.4 million in 1980, assisted in the resolution of specific, nationally important problems through the application of new scientific and technological knowledge. The objectives of the PFRA programs were to accelerate the application of basic research discoveries to the solution of practical problems; to facilitate the use of science as a working tool for problem resolution in the public and private sectors; and to enhance the capability and capacity of non-traditional research users to employ research results and methods. The criteria used in determining the suitability of projects for funding under the PFRA scheme included that they should generate recognizable and measurable public benefits; they should be multidisciplinary; they should contribute primarily to problem solution and not to the advancement of knowledge *per se*; they should provide research results that are important to the public and private sectors where disciplinary constraints, fragmentation, regulation or lack of market incentives inhibit research initiatives by others.

The Division of Intergovernmental Science and R & D Incentives, subsequently re-named the Division of Intergovernmental Science and Public Technology, had an Industry Program which encouraged university-industry co-operation and aided small businesses. The objectives of the Industry Program were to develop innovative mechanisms to couple the longer-range and more fundamental perspectives of academic scientists with the research needs of industry; to aggregate fundamental research interests across a given industry or technology for jointly-funded experimental projects in which a single firm could not ordinarily capture the benefits of the research; and to exploit the high incidence of technological innovation among small— and medium—sized firms. These various programs were a response to the priorities expressed by the legislature and, after 1979, to the President's statement on the stimulation of industrial innovation. Reorganizations have taken place within the Directorate to take account of these indicators. In 1979, the Applied Science and Research Applications Directorate was combined with Engineering to form the Engineering and Applied Science Directorate. In response to misgivings expressed in Congress, this was abolished in 1980 and two new Directorates created—the

Engineering and the Scientific, Technological and International Affairs Directorates. The latter has three main divisions — the Industrial Science and Technology Innovation Division, established to service President Carter's initiatives in the promotion of industrial innovation, the Intergovernmental and Public Service Science and Technology Division, and the Division of International Co-operative Scientific Activities. Under a continued Carter Administration, the budget of the Directorate would have grown from $56 million in 1981 to $80 million in 1982. The reality was quite different, however. The Reagan Administration considered many of the NSF's activities in the area of industrial innovation to have low priority and the Directorate's budget was slashed to $37 million in FY 1982. The budget of the Division of Industrial Science and Technology Innovations was cut from $46 million to $17 million and the new Office of Productivity, Technology and Innovation, established within the Department of Commerce, which was to support co-operative generic technology centres, was recommended for abolition. Tax incentives, instead of subsidies of the sort envisaged by the incentives programs, are intended to provide the private sector with the stimulus to invest in R & D.

In conclusion, the NSF has been less than successful in its endeavours to respond to political pronouncements and a general atmosphere within the legislature. Even if RANN was a valuable learning experience, the Foundation's experience with its successors shows that when priorities are set as a result of political considerations, the resulting programs are subject to the slings and arrows of political fortune.

The National Institutes of Health

The mission of the National Institutes of Health (NIH) is to advance the nation's capabilities for the diagnosis, treatment and prevention of disease. Administratively located within the Department of Health and Human Services, it had an estimated budget in FY 1981 of $3.9 billion and the budget request for 1982 was $4.2 billion. No formal statement of the principles followed by NIH in formulating and implementing national biomedical science policy exists but some general statements of the concepts that guide it have been made. NIH supports fundamental research to enlarge our understanding of the underlying biology of health and disease and it develops and conducts organized research activities aimed at specific disease or problem areas where the opportunities for a major advance seem clear and where circumstances and societal urgencies warrant an organized centrally managed effort. The most effective way that the organization believes a central body can broaden and exploit biomedical knowledge is by

- drawing on the best technical advice and evaluation regarding the quality of proposed projects, the state of the art, and significant opportunities and problems;
- stimulating those individuals with the most creative scientific minds and the most productive new ideas to do their best work by providing working arrangements and conditions that favour productivity;
- providing resources for directing new knowledge back into the research network and for making it widely available to the health services community.

The cornerstone of the NIH advisory system, regarded within the organization as a major contribution to the management of large federally-funded science programs, is the dual review process. This process is designed to keep a clear separation between decisions on the scientific merit of a proposed project and decisions on relevance to Institute programs and relative importance from the point of view of availability of Institute funds. This separation is achieved by assigning the decisions to two different sets of advisory committees. The scientific merit of all new research proposals is judged by specialized external initial review groups (IRG) or study sections. Program relevance is judged by an advisory council to the Director of the Institute in question. These councils are composed of knowledgeable laymen and scientists with recognized competence in the disease areas of the Institute concerned. The IRG decision on scientific merit is made first, and proposals recommended for approval are given a priority rating to guide the advisory council and Director. The council, assisted by Institute staff, assesses the relevance of the proposal to the Institute's program. It then recommends to the Director, in priority order, those proposals it considers relevant. In setting priorities, the councils generally give considerable weight to the IRG 'scientific merit' evaluation.

Operating as it does in the highly political area of health care, the NIH has, for many years, been acutely aware of the need to set priorities and explain these priorities to a public that, in this area of endeavour more than any other, expects 'the biggest bang for the buck' in research.[16] As noted in Chapter 2, a report on criteria for the allocation of resources to R & D was contracted by NIH to Professor R.A. Rettig and a subsequent NIH issue paper concluded that, as a result of this and other reviews of the literature, no theoretical basis for examining the appropriate level of research support to biomedical sciences or for determining the criteria upon which such support might be divided up was 'operationally adequate'. The practical consequences of this judgement for NIH were that budget recommendations had to be developed pragmatically, and in the development of NIH budgets three major considerations would have to be taken into account; the objectives of the program in question, the extent and emerging opportunities, and the relative and absolute limitations. The influence of Weinberg's criteria on the formulation of these considerations was evident; opportunities, for example, were explicitly related to Weinberg's intrinsic criteria.[17]

In the NIH's Forward Plan for FY 1976-80, Weinberg's influence was even clearer. In dealing with the question of central management versus decentralized initiatives the NIH administration came to the conclusion that the biomedical research enterprise meets Weinberg's criteria for a decentralized rather than a centralized approach, 'an approach that NIH has traditionally emphasized'. This approach, as it is understood within NIH, focuses on recruiting good scientists and providing them with support. If the scientists are sufficiently good then good results will be produced. The detailed research strategy develops after the fact and is the result of many small decisions taken by individuals rather than being laid out in advance by a research director.

NIH, therefore, justifies its approach in terms of Weinberg's criteria of scientific choice. A decentralized approach is preferred to a centralized one, and this, it is believed, illustrates a remarkable degree of self-discipline on the part of NIH science administrators since centralized programs are inevitably an exercise in power. However, despite its general belief in the desirability of a decentralized approach, the NIH has not hesitated to mount centrally-directed major programs. The National Cancer Program is a notable example.

In the light of this discussion, it would not be accurate to suggest that NIH actually uses Weinberg's criteria in allocating resources; it has simply noted that the research strategy that has traditionally been pursued by the agency is in accordance with them. In an attempt to refine the mechanisms by which resources are allocated and in response to requests from Congress to clarify its priorities, a study was carried out in 1975 that matched expenditures against mortality rates, days of hospitalization, visits to physicians and so on. Rank order correlations showed that the appropriations matched the various parameters quite well with the exception of cancer research which received twice as much money as heart disease research despite the fact that cancer kills about half the number of Americans annually as heart disease. The political nature of NIH funding is revealed in this particular case. The current thinking within NIH is summed up by Dr Solomon Schneyer of the Division of Program Analysis in the Office of the NIH Director:

> I don't know how much we should spend on biomedical research. This involves a question of social values. When we are dealing with billions of dollars of public funds it is difficult to see how this decision can be made by any formula. My own feeling is that we should spend what we can afford

and how much we can afford is a political question which will eventually have to be answered in the political arena.[18]

Summary

The R & D system in the United States can be described as pluralist and decentralized. No single department or agency is responsible for the government's research effort and the nation's science policy is the sum of all the science policies of these elements of the system. Apart from the checks and balances of the congressional budget process, there is no effective co-ordination or control. The first science policy model to be adopted in the United States after the Second World War approximated to what we have called the co-ordination system, with an independent advisory council and a federal co-ordinating council of senior departmental and agency officials. This machinery was abolished in 1973 and for three years there was very little scientific presence in the White House. A science policy machinery was restored in 1976, when the OSTP was created and an independent advisory council was added in 1982. It is clear that the influence of this science policy apparatus is strong and growing in the budget sequence.

National priorities in R & D can be detected in the way in which budget allocations made to the departments and agencies change over time. The way in which these priorities are arrived at has been described as an arcane decision-making and bargaining process embedded in a maze of executive and legislative systems. At the departmental level priorities are set internally. In the NSF, the agency that is responsible for the support of basic research, priority setting has been described in the following way: Political feasibility is the big thing—when there is a big earthquake the earthquake people get the funds. We worry less about statistics than about what will get through. Right now climate is a big area because of fears of drought and shortage but no one carefully analysed data before deciding to keep funds in that area. Sure, some data were collected but just to rationalize decisions that had already been made.[19]

CHAPTER 5

Science policy in the United Kingdom

The post-war period

The wartime mobilization of scientists and technologists in Britain had an enormous effect on the country's post-war science policies. At the operational level there was an expansion into nuclear, aviation, and electronics research. Advisory systems were established in the defence and civil sectors which, it was hoped, would steer the nation's growing R & D effort into areas of national concern. The Labour Government which took office in 1945 was committed to the relief of long-standing social inequities and to new programs of social and economic planning. It was convinced of the importance of R & D to the achievement of its goals but the possibilities for implementing a science policy that took into account these new social priorities were strictly limited. Britain's economic and political status had been seriously undermined by the war, and international trade and finance had to be considered in all domestic policy-making. The colonial legacy affected British defence policies for many years after the war and hindered economic co-operation with Europe. The 'special relationship' with the United States and the government's determination to possess an independent nuclear deterrent forced the British to undertake a massive rearmament program. The nuclear weapons program, its commitment to NATO and the defence of interests in the Near and Far East entailed an enormous diversion of resources and technological capabilities away from civilian industrial production, and Britain was launched into an arms race in advanced weaponry that she could not afford and that seriously distorted the R & D profile of the nation for many years and, to some extent, still does.

An attempt was made to improve the machinery for scientific advice and co-ordination with the creation, in 1947, of the Advisory Council for Scientific Policy (ACSP). The chairman, Sir Henry Tizard, was also the chairman of the new Defence Research Policy Committee (DRPC) and it seemed at first that this machinery would be sufficient for the co-ordination of the national R & D effort. The powers of ACSP were, however, severely limited. It had no executive responsibility, no power to allocate funds, and no power to initiate research or determine priorities in any way. When Tizard resigned in 1951 the ACSP and DRPC were completely separated again and there was even less co-ordination of civil and military R & D. British science policy immediately after the war has been summarized by Vig as follows:

> For the most part, the Labour Government's aspirations to redeploy science and technology for social and economic purposes came to naught. Civil research expenditure was increased significantly but pre-war institutional arrangements were largely consolidated in the post-war years. Science policy co-ordination was improved temporarily, but 'mission-oriented' programs in military and other departments soon overwhelmed general policy considerations. These were expanded in the years thereafter.[1]

In the expansion that followed, civil and military nuclear development heavily influenced university research priorities as well as governmental ones. High energy physics and related disciplines, especially, began to dominate university research expenditures. Nuclear energy and aviation absorbed at least half of all civil research funds through the 1950s in addition to most of the military R & D expenditure. Thus, the priorities of rearmament and weapons development led to a concentration on big science activities in the universities and in government. This, in turn, led to ongoing commitments that had to be met, making priority planning an impossibility for ACSP or any other sub-cabinet advisory body. Little change occurred in government scientific organizations during the lifetime of ACSP, and it made no impact whatsoever on the important matters of harnessing science and technology to increase national productivity and setting guidelines for economic policy. By 1963 a serious crisis had developed over government aid to science and technology. Public expenditure on R & D had been rising exponentially for some years, averaging 12-13 per cent growth per annum, but there was no effective machinery for co-ordinating this expenditure or for establishing budgetary priorities. The Department of Scientific and Industrial Research (DSIR) and ACSP recognized the need for priorities but declared themselves incapable of setting them. There was also some concern in DSIR and ACSP over the apparent failure of the large R & D investments to pay off in the industrial and economic area.

A Committee of Inquiry into civil science was set up in the early 1960s under the chairmanship of Sir Burke Trend and in its report to the government in 1964 it recommended that

- a new Ministry of Education and Science be established;
- the ACSP be abolished;

- a new advisory committee, the Council for Scientific Policy (CSP), be established to advise the Secretary of State for Education and Science on the distribution of funds among the Research Councils;
- the DSIR be abolished;
- three new Research Councils be established. The Science Research Council would take over the granting function of DSIR and appropriate research establishments, the Natural Environment Research Council would be responsible for the granting function of the old Nature Conservancy and its establishments, and an Industrial Research Development Authority would take over the DSIR's industrial R & D promotion activities and appropriate research establishments.

The new Labour administration which took office in 1964 implemented all of these proposals except one. Instead of an Industrial Research Development Authority a new Ministry was created, the Ministry of Technology, which was to fulfil Prime Minister Wilson's promise to remake Britain in the white heat of scientific revolution. A Central Advisory Council on Science and Technology, headed by a Chief Scientific Adviser, was established within the Cabinet Office to advise the government on the most effective strategy for the use and development of scientific and technological resources. With the change of government in 1970, there were again changes in the organization of the British Government's research system. The Ministry of Technology was abolished, a Department of Trade and Industry created in its stead, and a complete review of governmental R & D was carried out by Lord Rothschild, head of the so-called Cabinet think-tank, the Central Policy Review Staff. In his report, Rothschild recommended the adoption of a customer-contractor approach to applied R & D throughout the whole of the departmental research system.[2] His detailed recommendations, which the government accepted, were that

- applied research should be organized more widely in accordance with a customer-contractor principle under which the departmental 'customers', aided by their scientific advisers, would be responsible for defining their research requirements in terms of departmental objectives and for commissioning work from in-house or external 'contractors' who would advise on the feasibility of meeting particular requirements and put forward proposals for appropriate projects;
- in order to bring applied research in some areas covered by the Research Councils within the scope of the customer-contractor principle, some of the funds provided to the Councils should be transferred to customer departments;

- departments should become more closely associated with the framing of Research Councils' programs by having full members on the Councils, not 'assessors';

- the Council for Scientific Policy should be abolished and a new Advisory Board of the Research Councils established to advise the Secretary of State for Education and Science on the allocation of the Research Council science vote. This new Board would not be composed entirely of independent scientists, who would be in a minority;

- adequate central machinery should exist for ensuring co-operation and co-ordination between departments and for providing a concerted view on scientific and technological aspects of the government's policies, both domestic and international. Responsibility for this should be exercised by the Lord Privy Seal, a Minister without Portfolio.

From 1972 onwards the customer-contractor principle was introduced into the departments and these now commission the applied R & D they need from one of the Research Councils, an industrial firm or one of the department's own laboratories. Decisions on research requirements are related to departmental objectives and do not fit directly into any national science and technology policy. Indeed, Rothschild

Figure 5.1

Main elements of the U.K. Government R & D system

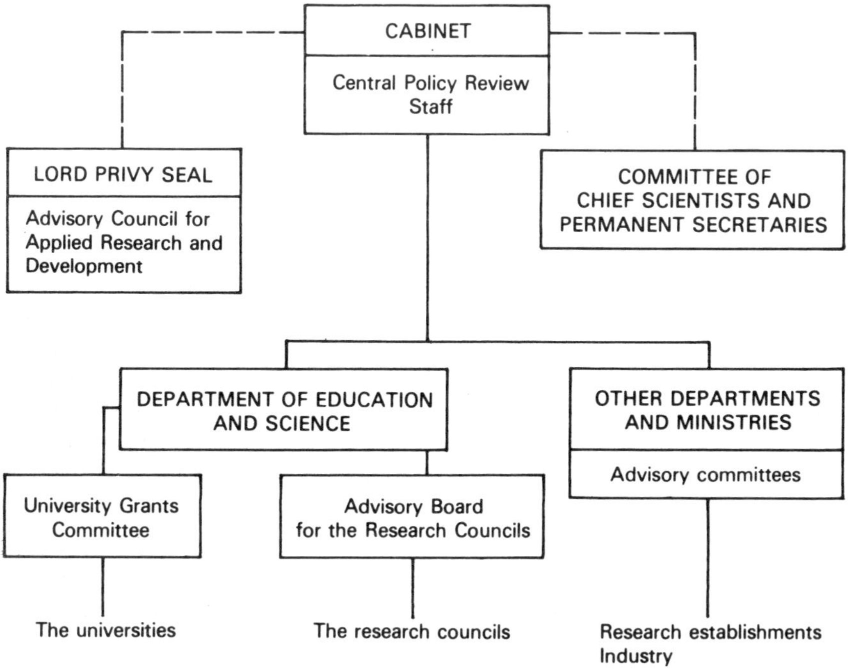

specifically stated in his report that any attempt to formulate overall objectives for a supposedly collective activity like R & D would only lead to confusion. In the new system the post of Chief Scientific Adviser to the government seemed unnecessary and when the then Adviser, Sir Alan Cottrell, retired from his post in 1974, no successor was appointed.

By 1976 it seemed that the new system was not providing the expected results and the government reviewed the arrangements that existed for the co-ordination of its research activities. Changes were made in the hope of achieving greater co-ordination and providing a forum of external advice to the government on applied R & D. A new advisory body, the Advisory Council for Applied Research and Development (ACARD), was established which, it was hoped, would provide a forum of external advice to the government on applied R & D. An overall co-ordinating committee of Chief Scientists and Permanent Secretaries was established, the office of the Chief Scientific Adviser to the government was abolished and a Chief Scientist was appointed to the Central Policy Review Staff. The new advisory and executive elements of the British research system are shown in Figure 5.1 and it can be seen that, on the surface at least, it appears to correspond to what we have described as a co-ordinated system.

The Cabinet Office

Under the general oversight of the Lord Privy Seal, the Cabinet Office has two functions. The first is a general supervision of the co-ordinating links between two or more departments; the second is more creative in that it involves both the provision of a forum in which important issues can be discussed and the initiation of reviews of subjects of interest to a wide range of departments in which there is a significant scientific and technological input. Thus, in its scientific role, the Cabinet Office

- provides a focal point for advising ministers or ensuring that advice reaches them on the science and technology aspects of the government's policies, both domestic and international;
- invites studies on certain trans-departmental issues involving R & D;
- monitors and advises on how the arrangements for the management of the government's R & D effort are working;
- assembles information to ensure that Parliament and the public have readily available sources of information about departmental R & D programs.

In order to ensure an overall review of R & D and other scientific matters at top official level, the Committee of Chief Scientists and Permanent Secretaries under the chairmanship of the Secretary of the Cabinet is responsible for seeing that scientific questions are brought before ministers collectively as appropriate and that scientific priorities reflect those of the government as a whole. It was originally intended in the Rothschild recommendations that this function would be performed by the Chief Scientific Adviser.

The Central Policy Review Staff (CPRS) within the Cabinet Office has responsibilities which cover the whole spectrum of government business but, from its inception, it has been interested in issues which have a scientific content, and it has always included a number of graduate scientists among its members. The CPRS assists the government in

- working out the implications of its basic strategy in specific areas;
- establishing the relative priority to be given to different elements of the government's program;
- identifying areas in which new options could be examined;
- ensuring that the underlying implications of the courses of action are fully analysed and considered.

Prior to the establishment of the office of Chief Scientist within the CPRS, concerted advice to ministers on the science and technology aspects of government policies was arranged through the Cabinet Office machinery at official level where the view of the departments and their Chief Scientists were co-ordinated — normally under the chairmanship of the Head of the Science and Technology Group. The initiative lay with one or more departments or it may have arisen from studies initiated by the Science and Technology Group itself. When the Chief Scientist was appointed to the CPRS the Science and Technology Group in the Cabinet Office was dissolved. The Chief Scientist works with other professional staff in a multidisciplinary team and is a member of the Committee of Chief Scientists and Permanent Secretaries and an 'assessor' on ACARD. There is considerable scope within the terms of his appointment for direct submission of advice to Cabinet on scientific and technological matters, and he has access to, and can comment on, all departmental reviews and reports prior to their submission to Cabinet. The role of the CPRS is not primarily one of co-ordination. Rather, it is one of asking fundamental or innovatory questions and of undertaking studies or projects that are best led from the centre.

The Advisory Board of the Research Councils

The Advisory Board of the Research Councils (ABRC) was set up in

1972 as successor to the Council for Scientific Policy to advise the Secretary of State for Education and Science on two matters; his responsibilities for civil science with particular reference to the Research Council system, its articulation with the universities and government departments, the support of postgraduate students and the proper balance between international and national scientific activity; and on the allocation of the science budget amongst the Research Councils and other bodies, taking into account funds paid to them by customer departments and the purposes to which such funds are devoted. The Board is also charged with the promotion of close liaison between Councils and the users of their research. The membership of the ABRC is quite different to that of its predecessor. Whereas the CSP was entirely composed of independent members, in the ABRC these are in a minority. The current membership includes the chairmen of all the Research Councils and the chairman of the University Grants Committee, the Chief Scientists of four departments and the CPRS, and seven independent members. In discharging its responsibilities on the more general matters listed in its terms of reference, the ABRC carries out major reviews and convenes standing committees and expert working groups to consider matters that have been brought to its attention or have been generated within the Board itself.

The most important function that the ABRC performs is the allocation of resources to the five Research Councils. In advising the Secretary of State on the distribution of the science budget amongst the Research Councils the Board uses a forward planning procedure that it calls the Forward Look. In the context of the science budget (the Research Councils' vote), forward planning has long been a standing feature of British science policy. The CSP conducted regular Forward Looks from about 1965 onwards, though these were not as rigorous as the ABRC procedure. Basically, the CSP Forward Looks involve adjustment of the Research Councils' proposals to a given growth rate, thereby pressing the Councils to reveal their uncommitted resources. The ABRC continued with this procedure in 1973 but it then reached the conclusion that the forward planning mechanism should be strengthened if the Board was to play an effective role in advising the Secretary of State. A new procedure was worked out, in collaboration with the Research Councils. The Councils are now asked to submit to the ABRC their own Forward Looks in a format called the Common Framework, which allows for comparisons of the policies and activities of five very different bodies.

In view of the built-in constraints that inevitably exist in any attempt to change the Research Councils' mainline expenditures, the ABRC recognized that an annual review could not constitute a major assessment of their main programs. It was considered, however, that there was room for manoeuvre at the margins and the Councils are asked to make their submissions at three assumption levels. At the upper level, an increase in funds at a particular percentage is assumed. At the main level, there is no increase from the previous year's allocation and, at the lower level, there is a reduction. Thus, each Council is asked to take as the main guideline the previous year's allocation suitably adjusted to take into account an increase or decrease in inflation and then to assess the effect on their programs of a decrease or increase of a certain percentage. In reaching their decisions on activities that would be increased or curtailed, Councils are encouraged to use a set of common science policy and management criteria that the ABRC has adopted as relevant to its overall aims. These criteria are set out in Table 5.1.

Table 5.1

ABRC criteria

Scientific policy criteria	Management criteria
• Excellence of study field: where benefits are attributable to a high proportion of the research being intrinsically of high intellectual value.	• Efficiency of operation: where improvements in organization and/or plant would lead to a general increase in efficiency.
• Excellence of the research workers: where benefits are attributable to the exceptional quality of the individuals or teams to be employed.	• Obsolescence: where the maintenance of a capability (at whatever level of activity) requires replacement within the Forward Look period of a major item of obsolescent plant or equipment.
• Pervasiveness of the activity: where benefits include the impetus to advances in other and related fields of science in addition to the primary field.	• Timing: where a start on a new or increased activity within the Forward Look period is critical if the expected benefits are not to be lost or much reduced.
• Social and/or economic importance: where expected benefits arise from the work being directed to supporting social or economic aims.	• Dependence on science budget support: where there is likely to be limited support, national or foreign, available for work related to the activity except the science budget.
• Significance for the training of scientific manpower: where benefits will include training and experience for scientific workers.	• Availability of scientific manpower: where an activity attracts priority by virtue of greater availability of scientific manpower for it (or its execution is constrained by lack of it).
• Educational importance: where benefits will include a contribution to education.	• Scope and limits of redeployment.
• Significance in maintaining national scientific prestige: where benefits will contribute to national reputation.	

In the first stage of the process, which takes place in January each year, the ABRC gives to the Council the provisional guideline for its Forward Look submission. This gives the Council the figures that it can expect for each of the next four years, and it is asked to prepare its submission using the guideline. In accordance with the Forward Look procedure the Research Council is also asked to say what it would do if it got x per cent more or less than the main guideline; x normally is two. It then sends the guidelines to its own Divisions and Boards and asks them to submit their estimates. In February the Divisions and Boards send their Forward Looks to the Research Council and it is here that the hard bargaining begins. In March the Research Council makes its decisions and the Boards and Divisions revise their parts of the Forward Look to take these decisions into account. A submission is then made to the ABRC, where it is defended by the head of the Research Council, assisted by whatever staff he may require, before an ad hoc committee of the ABRC. This committee includes the Chairman of the ABRC, as many independent members as can attend, and any of the *ex officio* members of those bodies whose Forward Looks are being examined. When all Forward Looks have been examined, the independent members formulate the advice to be sent to the Secretary of State. This advice, which is subject to the agreement of the ABRC as a whole, goes to the Secretary of State for inclusion in the Public Expenditure Survey.[3] The ABRC's advice is generally accepted though sometimes officials make small adjustments at the margins. In September the Research Councils receive the formal guideline decision from the Department of Education and Science and the budget is then published in the Public Expenditure White Paper. Decisions within the ABRC continue to be made by traditional bartering, as might be expected from its composition. Since the common criteria are not weighted in any way they serve only to ensure that the submissions from the Councils will have a certain uniformity and this may facilitate decision-making during the 'horse trading' within the Board. No formal mathematical techniques have been developed for decision-making, nor are such methods contemplated.

The Advisory Committee on Applied Research and Development (ACARD)

The Advisory Committee on Applied Research and Development (ACARD) was established in 1976 as a separate body to improve the interface between government and external organizations in the area of applied R & D. The ACARD complements the work of the ABRC but is distinct from it, and differs from it in several ways. Unlike the ABRC, it concentrates mainly on technology and the industrial application of science rather than on 'pure' science. It does not manage scientific institutions nor does it allocate resources. Its terms of reference are to advise ministers and to publish reports as necessary on:

- applied R & D in the United Kingdom and its deployment both in the public and private sectors in accordance with national needs;
- the articulation of this R & D with scientific research supported through the Department of Education and Science;

- future development and application of technology;
- the role of the United Kingdom in international collaboration in the field of applied R & D.

The Chairman of ACARD, the Lord Privy Seal, is a Minister without Portfolio and its members are from the non-government sector. The Head and the Chief Scientist of the Central Policy Review Staff, the Chief Scientific Adviser of the Ministry of Defence, and the Chief Scientists at the Departments of Energy, Industry and Environment are permanent Assessors. Other Chief Scientists attend by invitation as appropriate.

The ACARD convened for the first time in December 1976 and a number of working parties have been set up since that time. The first, set up in January 1978, had the following terms of reference:

- to project the effect of developments in semiconductor technology, including microprocessors, on United Kingdom industry over the next seven years;
- to identify industries that will be particularly affected;
- to consider the effects on products, production technology and design and development methods in these industries;
- to attempt to identify opportunities and dangers for United Kingdom industry from these developments.

In February 1978 the Prime Minister, in a speech to the Parliamentary and Scientific Committee commented on the work of the ACARD:

> The purpose of the Council is to advise the Government about the industrial significance and consequences of new scientific and technological advances in a period of rapid discovery and increasing complexity ...
> Technologically advanced industry has profound social and economic implications for our existing industrial structure and it is the task of government and Parliament to reconcile the pace of technological change with the inevitable social changes that follow, so that society itself is not disrupted. The three working parties (of the Council) are considering strategic problems of the greatest importance to our future.

This speech attracted a great deal of media attention and led directly to the making of 'The Chips are Down', an influential film made by the BBC on the revolution in microelectronics. In its report the working party on semiconductor technology recommended that government and industry should promote studies to give the earliest possible sensible warning of the changes in employment patterns that could be expected from the 'microprocessor revolution' and plan action to bring these changes about smoothly and in a socially acceptable way. The government acted on these recommendations and established an inquiry into unemployment and technological change within the Department of Employment which will be considered in Chapter 9. The other two working parties to which the Prime Minister referred in his speech were on the conditions for innovation in British industry in comparison with selected overseas countries and on the wider social consequences and social acceptability of technological changes, including the important effects on employment.

The Research Councils and Ministries

Since 1972 the Rothschild customer-contractor principle has been built progressively into the management of government-sponsored R & D. Funds were transferred from the Science Budget of the Department of Education and Science to the 'customer' departments so that they could commission appropriate research from the Agricultural, Medical and Natural Environment Research Councils. The Science Research Council (since 1981 the Science and Engineering Research Council) and the Social Science Research Council were exempted from compliance with the customer-contractor principle because the first was concerned with pure rather than applied research and the second was, at the time, very small. The total budgets of all the Research Councils for 1982-83 amounted to about £480 million. Arrangements were made, by amending Research Council Charters, to appoint departmental representatives as full members of the Research Councils. The general organizational pattern for the customer departments would be clearly identified central organizations responsible for the formulation of departmental research requirements and the administration of research contracts whether carried out in-house or by other departments, universities, industry or the Research Councils. The main intentions of the Rothschild changes as they affected the Research Councils and the work of the ABRC can be summarized as follows:

- the customer-contractor principle already in use in major areas of government R & D would be applied more widely;
- the Research Councils would be preserved within the Department of Education and Science and not transferred to the departments that would be expected to be their major customers;
- departments would be more closely associated with Research Councils by full representation on the Councils themselves;
- the ABRC would be given information about the size and nature of the work to be commissioned by customer departments and to take this into account when advising the Secretary of State on policy for the Science Budget;
- the ABRC's recommendations would be considered by the govern-

ment in planning future public expenditure totals for R & D;
- no conditions would be placed on the use of the money transferred to customer departments but it was expected that it would be spent to commission applied research work from the Research Councils;
- Councils would be responsible for the detailed management of the commissioned research. Partnerships and co-operation between departments and Research Councils would be an essential feature of the new approach. The support which the Research Councils gave to the universities would continue unimpeded.

In its second report, issued in 1976, the ABRC claimed that a good start had been made towards realizing Rothschild's aims. The customer-contractor principle was intended to give the government departments an increased influence over the pattern and direction of the Research Councils' work and to make it easier for the Councils to be informed of the departments' problems and to adjust their own research programs accordingly. The ABRC noted that there had been some redirection of Research Council efforts, though not on a large scale. It attributed this relatively slow redirection to the fact that departmental R & D requirements needed to be defined more explicitly so that the Research Councils could plan their programs in anticipation of needs. It saw benefits in the greater contact then existing between scientists inside and outside government departments as a result of the new machinery, but set against these benefits the considerable time and effort spent by scarce, highly paid scientific and administrative staff in establishing and operating the new commissioning procedures. Insufficient information about the revival or termination of existing research commissions or the development of new ones coming from the customers was also causing difficulties for the Councils' efforts to provide continuity of employment for research staff. In its third report, the ABRC added little to these general sentiments and advocated no major changes to the system.

In March 1979 a major review of the effects of the Rothschild system on the Research Councils was carried out. The review, which was presented to Parliament by the Lord Privy Seal, gave the impression that in general the new system was working satisfactorily in all areas. But the difficulties experienced by at least one Research Council in dealing with the customer-contractor principle would seem to cast some doubt on the value of this review. The Medical Research Council was violently opposed to the introduction of a customer-contractor relationship between it and the Department of Health and Social Security and to the removal of a part of its budget to the Department. Rothschild expressed the view that a good case could be made for transferring to the Department of Health and Social Security (DHSS) about one half of the allocation made to the MRC by the Department of Education and Science. This view was based on an examination of the subjects under investigation in the MRC at the time that were considered to be relevant to the Health Department's operations. Only a quarter of the annual income was, in fact, transferred. The department itself was unable to put forward ideas for new research on that scale and they found themselves reacting to proposals from the MRC. Up to 1976 the Council negotiated 142 'notional' commissions with the Health Department that were derived from the existing program of the Council and only one specific new commission was negotiated.

By 1977-78 the proportion of the MRC's budget contributed by the Health Departments had fallen to 21.6 per cent leaving the Council some £900 000 'short' in its expected appropriation, a circumstance that did not foster the development of a more harmonious relationship between the customer and contractor. The DHSS, for its own part, has had to build up the necessary expertise to discharge its own separate functions for developing health and social services research and has not felt justified in also developing expertise for a full commissioning role in the biomedical area because its research activities extend into other areas. Partly for this reason and partly because the MRC was already engaging in biomedical research work of importance to the DHSS, there was little change in the MRC's program as a result of the new arrangements. In 1977, the department proposed a simplified system to the Council in which they would set out each year a range of policy and service problems to which biomedical research might be applied. The Council would then suggest how their research programs might relate or be developed in response. The Council accepted these proposals and they came into operation in mid-1978. By 1981, however, the secretary of the MRC announced that the customer-contractor arrangement was being abandoned in the MRC and that the Council had regained control over that part of its budget which had previously been earmarked for DHSS 'contracts'.

Co-ordination

The government agencies involved in the funding and performance of R & D are the Ministry of Defence and the Departments of Energy, Industry, Environment and Transport, and Health and Social Security. The approximate 1981-82 budget of the Ministry of Defence was £1.8 billion and those of the departments totalled about £1.7 billion. No single Ministry or department is responsible for the government's civil R & D effort, though most of the funds are channelled through just two — the Department of Education and Science and the Department of Industry.

The elements of a co-ordinated science policy machinery are present in the United Kingdom but to describe it as such brings a negative official response. The official view on co-ordination is this:

> Whilst there is co-ordination of the United Kingdom policy implementation, it occurs much more at the working level than at a central steering co-ordination level. To this extent the United Kingdom system falls between Brickman's atomistic and low levels of co-ordination.[4]

It would appear, therefore, that although there is a high level co-ordinating committee (the Committee of Chief Scientists and Permanent Secretaries), a Cabinet Office Chief Scientist and an independent scientific advisory committee (ACARD), the level of co-ordination of the nation's R & D effort is quite low. The ministers and departments set their own priorities and, in accordance with the customer-contractor principle, commission the research they need. Intradepartmental rather than overall co-ordination would seem to be the preferred science policy mechanism.

CHAPTER 6

Science policy in Australia

The context

The Commonwealth of Australia is a federation of six states and two territories which, with an area of just under 7.7 million square kilometres, is almost as big as the United States. More than 60 per cent of the land surface is desert and much of the population of 15 million is concentrated in a crescent of fertile land lying along the eastern and south-eastern seaboards. The problems associated with the settlement and exploitation of such a vast, dry continent with soils of poor structure and low nutrient value ensured that early scientific work, and especially that funded by the colonial governments, would be concentrated on agriculture, geology and meteorology.

Among the responsibilities retained by the states after federation in 1901 were education and agricultural and mining development. The universities, the first of which was founded in the middle of the nineteenth century under state laws, continued to be funded by them after 1901 and, as a consequence, were unable to develop a strong research capacity; state governments simply did not provide sufficient funds for them to do so. In the agricultural sphere the states carried out no long-term research and no attempt was made to co-ordinate their research activities. Early in the new century the Commonwealth Government began to consider the question of scientific research and concluded that it had a national role to play. A Commonwealth Bureau of Meteorology was established in 1906, a Chemical Analysis Laboratory was set up within the Department of Defence in 1910 and, in 1916, a temporary Advisory Council of Science and Industry was formed to pave the way for a permanent Commonwealth Institute of Science and Industry. The Institute was established in 1920 and from then on development of the governmental R & D effort followed a familiar pattern. The ministries acquired the R & D capability that they needed in order to carry out their functions and the central government long-range research institution flourished. The universities built up their research capacity, slowly at first, but at an increasing pace in the 1960s

Table 6.1

Sector of performance	A$ million at current prices				
	1968-69	1973-74	1976-77	1978-79	1981-82
Commonwealth Government	117	208	326	353	540
State Government	41	76	132	157	211
Private enterprise	119	191	160	206	288 [a]
Higher education	89	174	244	326	440
Private non-profit	2	6	11	13	17
Total	368	655	873	1055	1496

[a] This figure is updated from ABS preliminary data. Cat. No. 8105.0.

Source: Department of Science and Technology, Science Statement 1982-83

Table 6.2

Source of funds	A$ million at current prices				
	1968-69	1973-74	1976-77	1978-79	1981-82
Commonwealth Government	173	383	579	695	998
State Government	61	67	117	148	200
Private enterprise	116	186	152	180	257
Other Australian	9	7	10	18	24
Overseas	10	11	14	13	17
Total GNERD	368	655	873	1054	1496
Commonwealth Government funding as % of GNERD	47	58	66	66	67
State Government funding as % of GNERD	17	10	13	14	13
Private enterprise funding as % of GNERD	32	28	17	17	17

Source: Department of Science and Technology, Science Statement 1982-83

when funding was first shared between the states and the Commonwealth, and in the 1970s, when responsibility for university recurrent funding was taken over completely by the Commonwealth.

A feature which distinguishes Australia's research system from most other OECD countries is the extent to which the funding and performance of R & D is dominated by government. In Table 6.1, the breakdown by sector of performance of the gross expenditure on R & D for 1968-69, 1973-74 and more recently is shown and in Table 6.2, the breakdown by source of funds for the performance of R & D for the same years is shown. The figures in these two tables demonstrate that the Commonwealth Government's role as both funder and performer of research is dominant and increasing and that there has been a drastic reduction in the R & D expenditure in the private sector. The reduction, which between 1973-74 and 1976-77 amounted to about 50 per cent, has led some commentators to the conclusion that Australia is undergoing a process of de-industrialization. Many industrialized countries have shown a decline in the proportion of their working populations employed in the manufacturing sector, especially during the 1970s. This trend is continuing, and is due to factors such as the persistent economic recession and technological change. What distinguishes de-industrialization from temporary effects arising out of economic recession or structural changes in industry through the use of more sophisticated technology is the total abandonment of certain industrial activities because of failure to compete effectively in the international market. On the basis of figures showing the decline in the manufacturing sector's contribution to the gross domestic product and exports (Table 6.3), the decline in manpower engaged in R & D and the decline in the level of its R & D expenditure, Johnston has concluded that Australia is de-industrializing, a process which, if allowed to continue, will 'take Australia back into the category of a Third World nation, fortunate in its resources but exploited by knowledge-, information- and capital-rich nations.'[1] If this is indeed the case, then the process will have been occurring during a period in which the country has been subjected to the most searching analysis of its R & D system and science policy machinery since federation.

Table 6.3
Contributions to GDP and exports by sector

Sector	1969-70		1977-78	
	GDP	Exports	GDP	Exports
Agriculture	7.1	54	4.4	45
Mining	2.8	21	7.2	28
Manufacturing	23	20	18.4	21
Other industries	10.7	5	6.5	6
Services	46.4	—	52.1	—

In the years following the Second World War, Australian manufacturing industry was working to an import replacement plan set out by the government in a 'White Paper' issued in 1945. Research and development did not have a high priority in this plan since it was based on borrowed product and process technology. The White Paper became the blueprint for the nation's industrial strategy over the next twenty-five years. Sheltering behind tariff barriers, import restrictions and quotas, and utilizing overseas technology, Australian manufacturing industry generally did not perceive a need to establish a strong R & D capability. In 1961, responding to continued criticism about this state of affairs, the Minister for Trade and Industry requested the Manufacturing Industries Advisory Council to examine the technical research needs of the small businessman. An interdepartmental committee was also established, and the outcome was the Industrial Research Grants Act of 1967. Under the Act, a company which spent more on R & D than it did in a base year could claim a grant covering a proportion of this expenditure. For increased expenditure up to $50 000 the grant was automatic but above that figure an Industrial Research and Development Grants Board exercised discretion. The Act was revised in 1976 because general grants unassessed by the Board increased to such an extent that no selective grants could be given and there was criticism within government that, as a consequence of this, it had no control over the projects funded. The revised Australian Industrial Research and Development Incentives Act introduced three types of grants. Commencement grants are made to companies which have not spent more than $250 000 on industrial R & D in the previous five years, or have received less than five annual grants for research support. These grants are designed to enable small companies, whose research experience is limited, to start work on projects with substantial R & D content. In 1981-82, approximately $10.5 million were set aside for such grants. Project grants provide the bulk of the government's direct assistance for industrial R & D. Up to $750 000 may be awarded to individual companies or groups of related companies in any one financial year for approved projects; in 1981-82, $32 million were allocated. Public interest projects are contracts made to industry on behalf of the Commonwealth Government for projects which normally involve the commercial development of the research results of government or private non-profit research establishments. The projects are of high priority in terms of government policy objectives and are expected to have substantial social and economic benefit. In 1982-83 $5 million were set aside for these projects.

The Commonwealth Government's current direct support for industrial R & D now amounts to about $50 million per annum. Direct support remains the government's preferred strategy for providing incentives to manufacturing industry to carry out its own R & D, despite the fact that various advisory bodies have suggested that it should consider other methods such as tax incentives and contracting out. The Australian Science and Technology Council (ASTEC), a

high level independent advisory body, has consistently recommended
that the government adopt a policy of contracting an increasing proportion of its R & D requirements to industry, establish a tax incentive
scheme for industrial R & D expenditure and increase the level of its
support for industrial R & D.[2] A committee of inquiry, the Study
Group on Structural Adjustment, made the same recommendations in
1979 and, like ASTEC, strongly advocated the establishment of an
Australian Innovation Authority, similar to Britain's National Research Development Corporation, to effect the transfer of research
results from the public to the private sector.[3] Opposition to such an
Authority by the Department of Finance, Treasury and the nation's
leading research organization, the Commonwealth Scientific and Industrial Research Organization (CSIRO), has continued to prevail.
Despite evidence that direct subsidies to industry for R & D are less
effective than tax incentives and other measures, the Government
refuses to introduce them. The level of procurement through contracting out of the government's R & D requirements to private industry
remains very small. In 1980-81, Commonwealth Government contracts
to private enterprise for scientific and technological activities
amounted to $28 million; of this, only $2.5 million was for R & D. Importantly, the direct subsidy technique does not appear to be working,
despite significant increases in the level of expenditure in recent years.
It was shown earlier that there has been a dramatic decline in the
manufacturing industry's R & D expenditure since 1973-74 and
although the figures for research expenditure levels in this sector show
a 5 per cent increase in real terms between 1976-77 and 1977-78, there
is still considerable concern about the obvious reluctance of Australian
industry to commit resources to this vital activity.

Towards a science policy machinery

Until 1972 no efforts were made to establish a science policy machinery
in Australia. There was no national scientific advisory body, no science
budget, no ministry with responsibility for science and technology
policy or for the development and co-ordination of the national research effort. It is true that in 1966 a Ministry for Education and
Science was created, but the science branch, though having formal
responsibility to work on science policy, concerned itself mainly with
the collection of statistics, not with the development of policy. Indeed,
the Liberal-Country Party coalition Government which ruled Australia
between 1949 and 1972 consistently denied that it was possible to
define a national science policy, a stance that removed the need to
consider the establishment of science policy machinery.

Post-war references to the need for a scientific policy are contained in the proceedings of a seminar on *Science in Australia*, held at the new Australian National University in 1951.[4] The imbalance in funding between CSIRO and the universities was a matter of great concern at the time, and science policy was seen by the participants in the seminar as a means of increasing the financial provision for pure science. Two recommendations were made. The first, recognizing the need for scientists to present a united front in their dealings with government, urged the establishment of an Academy of Science; the second suggested that in order to overcome CSIRO's monopoly of the provision of scientific advice to the government, an independent advisory body should be appointed. The first recommendation was soon translated into action and the Australian Academy of Science received its royal charter in 1952. The second was less easy to achieve and for the remainder of the 1950s there was little demand by scientists or politicians for the establishment of a science council or any of the other elements of a science policy machinery.

In the 1960s, however, the Academy of Science began to lobby for the creation of a body similar to the United States National Science Foundation to distribute funds for pure research, and for the establishment of a science council to advise government. Support for the establishment of a science policy machinery came from the chairmen of the CSIRO and the Australian Atomic Energy Commission as well, and in 1965 the leading politicians of all political parties joined in. Prime Minister Menzies in a statement to the House of Representatives announced

> . . . we have decided that there ought to be a body to advise the government on the most effective methods of co-ordinating, and achieving results from, expenditure on research through the universities, through the government's own agencies, and through any other bodies to which grants are made by the government. We will therefore study the various methods which have been developed for this purpose in overseas countries.[5]

Within months of this statement, the Australian Labor Party issued its own platform statement on science and technology. 'Australia,' it said, 'desperately needs national scientific policies which will embrace not only planning for scientific R & D, but also enable the results of scientific research in Australia and elsewhere to be applied in every aspect of Australia's industry and its culture.'[6] To administer the proposed science policy, the Party promised the creation of a Ministry of Science, an independent advisory Science Council with a rotating membership of senior academics, industrialists and government scientists, a Parliamentary Committee on Science and Technology, a National Science Foundation, and tax incentives to encourage Australian industry to undertake R & D.

The enthusiasm of the Liberal-Country Party coalition Government

for a science policy machinery soon evaporated and the demands of the
scientists were resisted. This is not to say that their demands were
ignored. Indeed, in reply to a document issued by the Academy of
Science calling for the establishment of a full-scale science policy
machinery the then Minister for Education and Science, Malcolm
Fraser, went to considerable lengths to explain his government's
position on the matter:

> The point I am trying to make, at the risk of unduly repeating criticism
> made overseas, is that there is no easy answer and that there is no simple
> solution to the problems of priority and that the machinery specially
> established to help determine these problems does not seem to have worked
> particularly well . . . We may then be wisest to continue our pragmatic
> evolutionary approach seeking advice from different people as different
> projects arise.[7]

The pragmatic Mr Fraser recognized that the government's science
policy was in an evolutionary phase, and he acknowledged the fact that
it might be persuaded to alter its policies. In April 1972, on the advice
of the Chairman of CSIRO and the Secretary of the Department of
Education and Science, the government did precisely that, and established the Advisory Council for Science and Technology to give co-
ordinated advice on actions and policies that would assist in the align-
ment of Australia's science and technology to national objectives. The
stage was now set for what was to be described as the longest running
show in Australian science.[8]

The Australian Science and Technology Council

Up until 1972 any routine advice that the Commonwealth Government
received concerning the adequacy, balance and direction of its R & D
effort came from the CSIRO and, to some extent from the Australian
Academy of Science, many of whose fellows are CSIRO scientists in any
case. For major policy initiatives, revisions or re-organizations, however, the government relied on the advice received from special committees of inquiry and it still does now, despite the existence of the
officially appointed independent advisory bodies.

The Australian Labor Party's platform statement on science and
technology, which was issued in 1964, committed it to the establishment of a Ministry of Science with no operational responsibilities
which, together with an independent advisory body, would formulate
policy recommendations for submission to Cabinet. The then leader of
the Party, Gough Whitlam, interpreted the platform somewhat differently, however. In 1970, two years before he became Prime
Minister, he said that the Minister for Science, together with an

advisory science council, 'would formulate policy recommendations for submission to Cabinet and would seek through the operational arm of his portfolio—the CSIRO and comparable organizations—and through the advisory council, the implementation of policy.'⁹ Whereas the platform foreshadowed the creation of what we would regard as a Ministry of Science Policy, the future Prime Minister's statement did not. The Public Service contingency plan, which was put into effect when the Labor Party won the 1972 election, followed Mr Whitlam's guideline, not that of the official Party platform, and the new Minister for Science, Mr W.L. Morrison, became the minister responsible for CSIRO and, in addition, acquired a number of laboratories which were made responsible to him through the Secretary of a new Department of Science.

One of the first executive decisions of the new Minister for Science was to abolish the previous government's Advisory Council for Science and Technology; some science policy analysts regard this decision as a blunder. Encel, for example, has said:

> This action was unnecessary and damaging, especially as it meant the lapse of over two years before the (new) Council was re-established in the shape of ASTEC. Politically, of course, abolition of the Council was a more spectacular decision, consistent with the general style of the Whitlam Government, whose career was punctuated by a series of spectacular and sometimes ill-conceived or even disastrous actions.¹⁰

In attempting to formulate the terms of reference, membership categories and methods of appointment of the Labor Government's own advisory body, Mr Morrison engaged personal advisers drawn from the ranks of the Labor Party. He then sought the advice of ministerial and departmental colleagues and of the scientific community by issuing a discussion document. Finally, he asked the OECD to conduct a review of scientific and technological activities in Australia which could become part of that Organization's series of examinations of the science policies of its member countries. The OECD examiners proposed a Canadian-style concerted action science policy system for Australia which had the following features:

- a Minister for Science without operational responsibilities;
- a Ministerial Committee chaired by the Minister for Science which would have a co-ordinating function and exercise its powers through control over a 'science budget';
- a powerful advisory council, possessing a small secretariat, which would assist the Ministerial Committee and report to the Prime Minister through the Minister for Science. Thus, 'on all matters of detail and substance the Ministerial Committee would be assisted by an Advisory Council for Scientific and Technology Policy, which

would prepare the work of the ministers and work out the details of the national science policy in consultation with appropriate experts.'[11]

The government's decisions on its science policy machinery, published in a White Paper in January 1975, bore a superficial resemblance to the OECD recommendations. The Ministry of Science would continue, a Ministerial Committee would be set up and an Australian Science and Technology Council (ASTEC) established. However, the Minister for Science would retain his operational responsibilities, exercised through statutory corporations and the Department of Science; the Department of Science would supplement and complement ASTEC, acting as its agent in obtaining and supplying data and in commissioning and conducting studies; ASTEC would be established by law but would report to the Ministerial Committee whose chairman would be the Prime Minister or his nominee. The Ministerial Committee would also be advised by interdepartmental committees on all matters referred to ASTEC and on matters considered inappropriate for reference to it. The structure of these interdepartmental committees was not revealed in the White Paper, but it is clear that by their retention the existing powers of the statutory corporations and departments in scientific and technological matters would be preserved and the powers of ASTEC curtailed. And since there was no mention in the White Paper of a science budget, the Ministerial Committee and ASTEC would be denied the prime co-ordinating instrument in any case. In effect, therefore, the Labor administration rejected the fundamental recommendations of the OECD examiners, as did Mr Fraser, the then Shadow Minister for Science, who dismissed them because 'examiners from overseas might not necessarily be able to make the best judgments about these matters.'[12]

An interim ASTEC was duly appointed and was promptly ignored by the government. Decisions which strongly affected the conduct of R & D in Australia were taken without reference to the new advisory body. For example, in June 1975, coinciding with a change of Minister for Science, the decision was taken to remove the Minerals Research Laboratories from the CSIRO to the Department of Minerals and Energy and the necessary administrative orders were issued. This was an almost unprecedented interference by an Australian Government into the affairs of a statutory scientific corporation and evoked howls of protest from the CSIRO and the scientific community in general. The compromise that was eventually worked out owed little to the influence of the interim ASTEC. The advisory body was more successful, however, in persuading the government to reverse a decision in August 1975, drastically reducing the amount of money allocated to university research workers under the peer-adjudicated research grant schemes operated by the Australian Research Grants Committee and the National Health and Medical Research Council. The budgets were increased after representations by the interim ASTEC and others. In the same month the Royal Commission on Australian Government Administration, which had been set up in 1974 to conduct a searching inquiry into the Commonwealth Public Service, established a science task force 'to examine the conduct and co-ordination of scientific work carried out, financed and/or supported by the government'. The interim ASTEC was neither consulted by the Royal Commission about the appointment of the task force and its terms of reference, nor by the science task force itself during its deliberations. The task force report, referred to in Chapter 3, was issued in November 1975, some two weeks before the government fell. Briefly, it recommended the continuation of ASTEC as a statutory advisory body reporting to the Prime Minister or a supra-departmental minister assisting the Prime Minister and abolition of the Science portfolio and the Department of Science.

The Labor administration was dismissed by the Governor-General in November 1975 because of the failure of Supply Bills to pass in the Upper House of Parliament and a caretaker government, led by Malcolm Fraser, took office. This government soon issued a statement on its science and technology policy which described the Department of Science as the scientific nerve centre of the Australian community, in apparent rejection of the science task force's recommendation for its abolition. ASTEC's role was considerably downgraded in this policy statement but a subsequent press release reverted to the rhetoric of previous Liberal Party platforms on science and technology policy. ASTEC, in a new Liberal-Country Party coalition Government, would have the highest status and the greatest independence, and would report directly to the Prime Minister on the whole range of scientific matters dealt with by the government. It is claimed by the task force convener, without much in the way of proof, that in restoring ASTEC to its former position as principal scientific advisory body the caretaker government was influenced by the science task force report. But it is difficult to accept that any government could have been swayed by the arguments put forward by the science task force in support of its recommendations.[13] When a newly-elected Liberal-Country Party coalition Government took office in December 1975, it soon became clear that, despite the pre-election rhetoric, ASTEC's future was by no means secure. Six of the Labor Government's nominees were dismissed and a small advisory group was established to make recommendations about the future of the Council. In April 1976, as a result of this group's recommendations, a second interim ASTEC was convened and asked 'to prepare a definitive report to assist the government in its decisions on the long-term future of ASTEC'. The recommendations of the second interim ASTEC were accepted by the government and

Australia now has a permanent advisory council established by law and reporting to the Prime Minister on

- the advancement of scientific knowledge;
- the development and application of science and technology to the furtherance of the national well-being;
- the adequacy, effectiveness and overall balance of scientific and technological activities in Australia;
- the identification and support of new ideas in science and technology likely to be of national importance;
- the practical development and application of scientific discoveries;
- the fostering of scientific and technological innovation in industry;
- the means of improving efficiency in the use of resources by the application of science and technology.

A full-time secretariat, administratively located within the Department of the Prime Minister and Cabinet, services the Council and its subcommittees, and carries out background and other studies appropriate to its work. Between 1977 and 1982, 17 reports containing numerous recommendations were submitted to the Prime Minister either at his request or arising out of ASTEC's own background studies. In the budget cycle ASTEC reviews new policy proposals and ongoing programs of the departments and agencies and is a source of confidential advice on these to Cabinet. The R & D system now in place is shown in Figure 6.1.

The Ministries

The annual *Science and Technology Statement*, which is issued by the Department of Science and Technology, listed 23 ministries in 1981-82 that performed or funded R & D, but only seven spent more than $5 million. The budgets of the seven are shown in Table 6.4. From this table it can be seen that the Ministries of Science and Technology and Defence taken together account for more than 70 per cent of the Commonwealth Government's entire expenditure on R & D.

The government's R & D effort is carried out either in statutory corporations or in departmental laboratories. The statutory corporations are established by law and report directly to a designated minister, not to the permanent head of an administrative department. This organizational arrangement for government science derives from the British experience and is in accordance with the principles laid down by Lord Haldane in 1918, when the research council system was created, becoming the model for developments in Canada, South Africa, India,

Table 6.4

Ministry/Department	1981-82 R & D expenditure A$ million
Science and Technology	371.5
Defence	118.3
Education	96.2
Resources and Energy	45.9
Health	38.9
Primary Industry	19.8
Foreign Affairs	17.7

Ceylon and Australia. CSIRO was the first scientific statutory corporation to be established and, as figure 6.1 shows, there are now five. Statutory status gives these organizations a considerable degree of autonomy in their everyday operations, an autonomy which is in harmony with the general belief in the scientific community that scientific research is best conducted in a different and freer atmosphere than that normally found in a government department operating under Public Service conditions. Advantages include the freedom from pressures of partisan policies with the consequent ability to concentrate on long-term projects; the creation of a personnel structure appropriate to its activities and free from the rigidities of ordinary Public Service rules; the ability to enlist the help of advisory boards of persons possessing particular qualities and experience; and the freedom to establish financial and accounting practices which meet the need for public accountability with the minimum of bureaucratic control.

In contrast to the statutory corporations, in which research of a more general nature is carried out, the departmental laboratories engage in R & D which is usually relevant to their immediate or medium term needs. The directors of the laboratories report to the Secretaries of the appropriate Departments of State, staff are engaged under terms and conditions laid down by the Public Service Board, promotion to more senior positions is generally controlled by the existence of vacancies at the higher levels rather than on merit alone, and budget requests are made on a 'line item' rather than 'single line' basis. Departmental research scientists have consistently lobbied for statutory status and the Science Task Force of the Royal Commission on Australian Government Administration recommended that most of the departmental laboratories, including the Defence Science and Technology Organization, be designated statutory corporations. This task force had originally been set up by the Royal Commission because of the deterioration of morale and the widespread sense of frustration evident in the submissions it had received from scientists employed in the departmental laboratories. Accepting this rather gloomy view of the departmental research system, the task force identified the problems as having their origins in the excessive uniformity, rigidity and centralization associated with departmental organization. Virtually all of the problems would be eased, if not wholly eliminated, by the introduction of the institutional arrangement which confers as much autonomy as possible as far down in the hierarchy as possible. Its view was that 'creative, productive science depends on the autonomous operation of self-imposed values and controls and it is ultimately self-defeating for a society or government to erode the autonomy of the scientific community.'[14] This argument was supported by the theories of Robert Merton, Michael Polanyi and Sir Karl Popper, which were in accordance with the aggregated experience of the task force. Merton's formulation of the scientific ethos and the norms of science, in particular, provided them with the link they needed between autonomy and the performance of creative, productive science. Thus:

> Each of these four elements (of the scientific ethos) can be preserved only if the scientific community can maintain its autonomy. Where any of these elements is lost scientific work loses effectiveness and integrity.[15]

Although it is never explicitly stated in its report, it appears that it is the freedom to set the agenda for research that the task force demanded for almost the whole of the government scientific service, the freedom to decide not only the solubility of problems, an uncontroversial and readily accepted right, but also the freedom to decide on what problems will be tackled. There is, of course, no justification for the assertion that there is a necessary connection between creative, productive science and the freedom to decide on the problems to be tackled. If we accepted that proposition we would have to conclude that most scientists are hacks, for the majority do not enjoy work conditions that allow full adherence to the Mertonian norms, nor do they have complete freedom to choose their own research problems. These conditions are usually found only in research organizations that are partly or wholly devoted to the pursuit of pure research and the demand for such freedom in government establishments usually evokes a hostile response from the non-scientific community. Since the publication of the Science Task Force's report, the government has consistently refused to re-designate departmental laboratories as statutory corporations, despite representations made by ASTEC on behalf of the Bureau of Mineral Resources, Geology and Geophysics[16] and committees of inquiry into the Bureau of Meteorology[17] and the Defence Science and Technology Organization (DSTO)[18]. The DSTO external committee used the arguments of the Science Task Force to no avail; the DSTO remains within the Department of Defence and the number of scientific statutory corporations is still five. Three of these, the CSIRO, the Australian Institute of Marine Science and the Anglo-

Australian Telescope Board, are within the Ministry of Science and Technology; the Australian Atomic Energy Commission is located in the Ministry of Resources and Energy and the Commonwealth Serum Laboratories are within the Ministry of Health. Some of these ministries have their own advisory bodies to assist them in the development of sectoral policies and in the allocation of funds to outside agencies. The Ministry of Resources and Energy, for example, has a National Energy Advisory Committee, formed in 1978, which advises the Minister on all matters relating to energy, including the balance of resources for research relating to the development of Australia's energy resources, and developments in technology both in Australia and overseas relevant to energy production and use. In addition, there is a National Energy Research, Development and Demonstration Council, also established in 1978, which advises the Minister on co-ordination of the government's energy research effort and on the disbursement of grants to universities and other outside agencies for energy research development and demonstration.

The Ministry of Science and Technology

In 1972 the incoming Labor administration fulfilled one of the promises that it made in its election platform and established a Ministry for Science. The new minister, Mr Morrison, was given responsibility for three statutory corporations, and a number of departmental laboratories which had previously been within the old Departments of the Interior and Supply. The Minister's department, the Department of Science, was responsible for the administration of the non-statutory elements of his portfolio and its tasks included the provision of secretariat facilities to various research funding and other committees. A policy division within the department was responsible for the development of policies and proposals, and providing advice in respect of the government's involvement and support for scientific and technological R & D in Australia. When the new Minister abolished the independent Advisory Council for Science and Technology, the Department of Science, through its policy division, might have become the most powerful advisory body on science and technology in Australia. The customary pattern for the development of science policy machineries in other countries has been that an independent advisory body is initially created, which plays an important and influential role in the scientific affairs of its government. Then, as a result of the heightened interest in science and technology which the advisory council creates within government and the increased involvement of the bureaucracy in science policy matters arising out of its servicing function to the advisory body, the influence of the independent body begins to decline. This has occurred in Belgium, Canada, France, the Federal Republic of Germany, The Netherlands, Sweden, Switzerland and, to a lesser extent, in Japan. There are a number of reasons for this not happening in Australia. Mr Morrison was the most junior minister and did not hold Cabinet rank; he was also Minister Assisting the Minister for Foreign Affairs and had oversight of the preparation for granting independence to Papua New Guinea. There was a lack of harmony between the new Minister and his senior departmental officials and he preferred to appoint his own science policy advisers from outside the department in his first years of office. He was also influenced by CSIRO and the Academy of Science,[19] organizations that would not be expected to promote the development of a powerful Department of Science. In a speech given to CSIRO Chiefs of Division in 1974, Mr Morrison made it clear that he received precious little help from the scientists in his efforts to enumerate a science policy. 'From the comments I received,' he said, 'a science policy to them meant money for scientists', a remark which echoed the famous statement made by Liberal Prime Minister Gorton in 1968:

> I don't know what a science policy is. The critics want an overall advisory committee to allocate funds, but I don't see the need for an advisory body. These committees are only a group of individuals pushing the barrow for their own disciplines.[20]

In mid-1975, Mr Morrison was promoted to the Defence portfolio and the Ministry became the Ministry of Science and Consumer Affairs led by Mr Clyde Cameron, who suffered a demotion from the higher ranking Labour and Immigration portfolio in the process. In November 1975 the Science Task Force issued its report, which recommended abolition of the portfolio and re-allocation of its functions to other ministries. Centralization was identified by the task force as the prime, almost unique source of the malaise in governmental research institutions and, although it never offered a definition of centralism, it is clear that this meant, for them, two things. These were the science policy machinery which most of the world's developed countries had adopted 'to generate and administer national science policy' and the departmental organizational form for governmental R & D. The task force's recommendations on the latter were referred to earlier. As for the former, it was logical for the task force to recommend the elimination of all traces of the centralized science policy machine, and the Ministry of Science, though not ASTEC, would have to go. Whereas the concept of ASTEC was 'the rational expression of the recognition of the wide dispersal of science and technology in the government'[21] the concept of a special science portfolio was not, as this tended to concentrate scientific activities in one ministry, perpetuating the notion of science as an end in itself rather than science as a means of

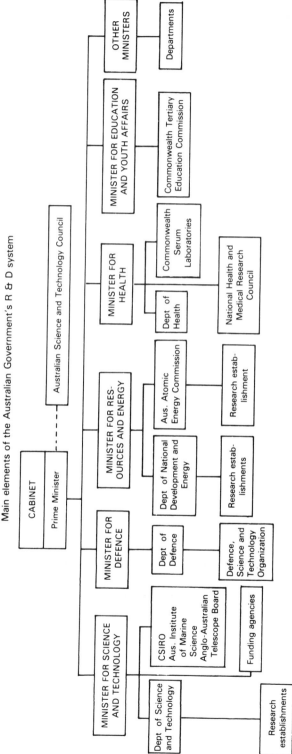

FIGURE 6.1:

Main elements of the Australian Government's R & D system

serving the objectives of other ministries. Had the task force suggested any credible way in which these other ministries might have been persuaded to use the research produced in the departmental laboratories 'liberated' by the abolition of the Ministry of Science and Consumer Affairs, these theoretical justifications might have had more force. A science adviser stationed in each department with a major scientific or technological component in its work would, apparently, accomplish the necessary technology transfer and all the autonomous statutory corporations would be on the alert to identify non-trivial societal needs and problems and extend their work in these directions.

The new Liberal-Country Party coalition Government clearly did not share the task force's view and the ministry was not abolished. In 1978 it became the Ministry of Science and the Environment and is now the Ministry of Science and Technology. The main areas of responsibility of the Ministry are shown in Figure 6.1. Two elements of the Ministry's work will be discussed—the Department of Science and Technology and the CSIRO.

The Department of Science and Technology

The functions of the Department of Science and Technology are to assist in the formulation and implementation of government policy for science, technology and productivity, and to provide scientific and technological services. The latter function is executed through the department's three operational divisions—the Bureau of Meteorology, the Antarctic Division and the Operations and Management Division. The policy division is responsible for the development of policy proposals and advice relating to the government's involvement in the R & D system. But for a number of reasons the provision of advice on the overall development of the government's research effort has not yet been an important aspect of the department's work. Factors which may have contributed to the low profile that the policy division has adopted include the lack of harmony between the first Minister and his department, the changes in the department's administrative responsibilities which are reflected in the numerous changes of name, uncertainty concerning the future of the department which persisted from 1975 until relatively recently, some hostility from the scientific community which continually fears 'bureaucratic control' over its activities, and the inability of the Minister to effect changes in the government's research system that the policy division might advocate because he was given no overall co-ordinating role. Though the Minister still has no such role, creation of the new Department of Science and Technology in 1980, with the transfer from another department of a new division of productivity development, and expansion of the role of the policy division to include the development of policy relating to productivity, has enabled the department to reassess and develop its role in the overall policy advisory arena. Recent important innovations include the production of an annual *Science and Technology Statement* which provides an excellent analytical breakdown of the ministries' outlays on science and technology, including R & D, and a Directory of Science and Technology in the Commonwealth Sector, both of which are important basic sources of information about the government's R & D effort. Seminars and workshops have been organized on science policy, technology transfer, stimulation of industrial innovation and technology assessment. The policy division has issued an important document on the bases of science and technology policy which analyses both government and opposition policies as they relate to fundamental science policy issues and seeks to define the ministry's role in the light of these policies.[22] The document proposed that, since science and technology are now combined for the first time in a ministerial portfolio, the ministry is well placed to become an agent of the integration of science and technology with social and economic objectives, a prime focus of interest in other OECD countries:

> Such a focus on the utilization of science and technology means that questions of integration and co-ordination of science and technology related policies and programs throughout government must be highlighted. The efficient integration of science and technology with social and economic policies is a challenge. This study suggests that it is a key theme of the new ministry in its development of effective policies to maximize the return on the substantial public investment in science and technology.[23]

Speculating on the type of science policy machinery that would enable the ministry to fulfil this integrated function, the paper advocates a co-ordination model — with emphasis on inter-ministry links and a central advisory body which, in the case of Australia, would be ASTEC. A four-point plan was put forward in the paper which, if implemented and suitably co-ordinated, would ensure the effectiveness of the nation's R & D effort. The plan, drawn from one of the Minister's statements was

- to develop and maintain a national capability in basic research which would provide a store of knowledge and expertise for the future;
- to develop a high-level capability in applied research which could provide the basis for not only capitalizing on domestic research activity but also using the results of international work;
- to transfer research achievements into the market place;
- to minimize undesirable social and environmental impacts of technological change.[24]

The paper had to acknowledge that the research appropriate to the first two points was outside the Department's sphere of influence. Most basic research is carried out either in the universities or CSIRO and although the Department of Science and Technology provides secretariats for some of the granting agencies that provide research funds to the universities, it has no control over the allocations. The same is true of CSIRO, which carries out a significant amount of pure research and most of the government's strategic research. Though the paper bravely suggests that there is a clear need for co-ordination in these and other areas of the four-point plan, the extent to which this could be done using the fairly loose co-ordination model that it advocates is problematic. The universities retain, and would be expected to defend, their traditional autonomy; the various independent funding bodies and the CSIRO, too, would undoubtedly resist any attempts to interfere with their own traditional autonomy.

The Commonwealth Scientific and Industrial Research Organization (CSIRO)

The CSIRO occupies an unchallenged position at the top of the Australian research system. It is the largest research organization in the country, employing about 7000 people, about 2000 of whom are research scientists. The work of the Organization is carried out in 39 Divisions and units, and these are organized into five Institutes. CSIRO's budget for the 1982-83 financial year is $313 million, about 35 per cent the Commonwealth Government's expenditure on R & D.

An Institute of Science and Industry was established by the Commonwealth Government in 1920 following the recommendations of an advisory committee convened in 1916 by Prime Minister Hughes. Rigid parliamentary control over its inadequate funds and federal-state rivalries reduced the Institute to the level of an inefficient advisory body to the Parliament. By 1923 it was clear that the Director could not carry out the main function that had been assigned to him, 'the initiation and carrying out of scientific researches in connection with, or for the promotion of, primary or secondary industries in the Commonwealth.'[25] In that year Sir Frank Heath, Secretary of the British DSIR, was asked by S.M. Bruce, Hughes' successor, to advise the Commonwealth Government on the reorganization of the Institute. In 1926, as a result of Heath's recommendations, the Council for Scientific and Industrial Research (CSIR) was established as a statutory corporation under the Science and Industry Act (1926) to advise the responsible minister on its policy and work, on the funds required for carrying out this work and on the allocation of funds made available by the government.

The new Council's functions were quite similar to those given to the Director of the Institute in the previous legislation and in addition State committees were established in order to advise the Council 'on its general business and any particular matter of investigation and research'. £250 000 was granted to the Council from Consolidated Revenue. By 1939 CSIR was the dominant force in Australian science. In the number of highly qualified research scientists employed, in the quality and sophistication of its research facilities and in the diversity of its research programs it was second to none. It had acquired a near monopoly in the performance of basic research, having outshone the universities across a wide spectrum of scientific research activity.

Most of the early pure science undertaken in Australia concentrated, according to Mellor, on geology and natural history.[26] A few enthusiasts in the universities performed a small amount of research in the physical sciences up to the beginning of the Second World War, hampered by isolation from the main centres of activity in Europe and North America and by lack of financial support from state governments. The statute under which CSIR was established seemed to imply that the universities were unable to meet the country's needs for research or for scientific research workers. The section of the original Institute of Science and Industry Act, which empowered that body to establish and award industrial research studentships and fellowships was amended to include the training of research workers as well as the establishment of industrial research studentships.

Two courses of action were open to CSIR when it was incorporated under its Act. It could have put its weight behind the universities, which its charter allowed it to do, building up research units within them and assisting them to develop strong postgraduate research schools; or it could have established its own research units to carry out research of all kinds appropriate to the needs of the agricultural and pastoral industries, leaving the universities to struggle on as best they could. The Council chose the latter course, sacrificing principle to expediency.[27] Reflecting on the fate of its predecessor, the Council was mindful of the need to achieve positive results as rapidly as possible in order to win the confidence of the Parliament. In his parliamentary speeches introducing the Bill, Prime Minister Bruce advocated the first option, but his actions belied his words, as the chroniclers of CSIR have indicated:

> The Prime Minister's actions in reorganizing the old Institute, giving it money and power to work under the new Act, making provision for advanced training of research scientists, and above all his selection of (Sir George) Julius and (Sir David) Rivett to be Chairman and Chief Executive Officer, respectively, led inevitably to the full development of a research organization. It seems unlikely that Rivett would have accepted the Chief Executive position in CSIR and left the Chair of Chemistry at Melbourne

unless he had believed from the beginning that the Commonwealth would develop its own research teams.[28]

Later on, when CSIR moved strongly into the physical and engineering sciences, the lack of postgraduate facilities in the universities ensured that they would be powerless to affect the dominance which the Council subsequently achieved in these areas too.

The Science and Industry Act of 1926 spelled out the functions which the Council would have but, by 1948, some confusion had arisen because these had reached beyond the statutory limits. The extension of its activities into the defence area was of particular concern to the Council and to Parliament. Until the 1930s, it only engaged in research relevant to manufacturing industry to a very limited degree. In 1936 the government announced its intention of encouraging CSIR to fulfil its obligations to the secondary sector. As a result of a report, a National Standards Laboratory and a Division of Aeronautics were established. During the Second World War CSIR made distinguished contributions in the areas of aeronautics, industrial chemicals and radar. After the war, Sir David Rivett, by then Chairman of CSIR, expressed the view that his organization should confine itself to civil research as the Act required and that all state defence work should be carried out within a separate organization. Parliament was also of that view, but for different reasons. In the anti-communist atmosphere of the time allegations were being made by the Opposition that CSIR and the universities were employing known communists and foreign agents and that, because of this, Australia was being denied secret information on atomic energy by the United States.[29] Since CSIR scientists were not subject to the provisions of the Commonwealth Public Service Act, and were appointed directly by the Minister on the advice of the Executive, it was widely felt that sufficient control over such appointments could not be exercised by Parliament or the bureaucracy in order to eliminate the possibility of entry by subversive elements. To resolve the situation, in 1948 the Cabinet invited the Chairman of the Public Service Board, Mr W. Dunk, and the Director-General of the Department of Post-War Reconstruction, Dr H.C. Coombs, to report on the types of organization suitable for the activities of the CSIR. As a result of their recommendations, CSIR became the Commonwealth Scientific and Industrial Research Organization. The governing body was to be an Executive of five persons, which had the power to appoint officers and other employees. In performing this function, however, it determined terms and conditions of appointment and these were then approved by the Public Service Board. The Division of Aeronautics was transferred to the Department of Supply under the provisions of the Public Service Board No. 2 Act, which gave the Governor-General power to declare any work being performed under the statutory corporation to be work which should be performed in a department of state, irrespective of its connection with defence matters.

These changes could have seriously affected the autonomy of CSIRO but in the event the new organization went from strength to strength, losing none of its real autonomy and continuing to dominate the scientific research scene in Australia. Its scientific officers, recruited worldwide, are able to engage in research relatively unhindered by other duties, and advancement within the organization generally rests on the excellence of the individual's contributions to the primary literature.

There is no doubt that CSIRO's reputation as a superior research organization is well deserved, secure and international, but it is sometimes suggested that this reputation was acquired at the expense of the universities which, until the mid-1960s, were poorly endowed. When the first universities were founded in the middle of the nineteenth century, they were planned on a liberal scale and were thought to be appropriate centres of science and scholarship. However, they were given inadequate funds for them to establish strong postgraduate schools in science and they tended to concentrate instead on the arts and professions.[30]

CSIRO AND THE UNIVERSITIES

If the historians of CSIR's early years could describe its relationship with the states as a 'spirit of sweet reasonableness' the same cannot be said for its relationship, and that of its successor, with the universities. In 1927, lines of demarcation in scientific research were drawn between CSIRO and the state Departments of Agriculture. Some of the basis of CSIR's and CSIRO's involvement in basic research, both pure and strategic, are to be found in this agreement:

> Investigations which are of a more or less fundamental character and which are national in scope should be conducted by the Commonwealth, while problems of more or less local character and which involve the application of knowledge should be undertaken by the States.[31]

No such 'gentleman's agreement' was sought by CSIR with the universities. The policy of the Council was, right from the beginning, to encourage its scientists to attack their problems on as fundamental a level as possible.[32] The 'encroachment' of CSIR into the traditional preserves of the universities was both easy and deliberate and, as the following statement by a member of the first Executive of the new CSIRO shows, was often expressed in a way that could not have improved the relationship between them.

> The 'scientific' laboratories have been deliberately created to fill recognizable gaps in Australian science. The Divisions of Meteorological Physics

and Chemical Physics were, at the time of their creation, pioneers in Australia in their own field. The cost of such research groups and even more so of the Radiophysics Group working in radioastronomy was, until quite recently, far beyond the purse of any university.[33]

The universities have long regarded themselves as the Cinderellas of the Australian scientific research scene. It is only in recent years that the funding and facilities for university scientific research have begun to approach international standards, or even the standards of CSIRO. Until 1948, when the Ph.D. degree was established in Australian universities, talented young graduates had to go abroad to obtain post-graduate qualifications. The first glimmer of hope that the parlous situation of the universities was being recognized at government level was the establishment of the Australian National University in Canberra — an institution which was to be concerned solely with postgraduate research and funded entirely by the Commonwealth Government. At the time, founding an entirely new postgraduate university, instead of increasing financial support for the existing institutions, did not reflect a lack of confidence in the universities; rather, it could be considered a reflection of the states' rights issue, which essentially proscribed any interference by the Commonwealth in areas in which the states had residual powers. However, the problems encountered by the states arising out of the great post-war expansion of student numbers were so great that by 1974 they appeared to be more than happy for the Commonwealth Government to take over complete responsibility for university funding.

Two events were of enormous significance in bringing about the change in fortunes of the universities. First, as a result of the recommendations of the 1957 Committee of Inquiry into Australian Universities, chaired by Sir Keith Murray, the Australian Universities Commission (AUC) was established on the pattern of the British University Grants Committee.[34] The AUC was abolished in 1976 and replaced by a Commonwealth Tertiary Education Commission with three Councils, one of which, the Universities Council, is responsible for advising the Commission on the funding requirements of the eighteen state universities. The universities' recurrent grants now include a special research component which provides floor support for academic staff to enable them to engage in research. In 1980, $72 million were distributed to the universities under this heading. Second, resulting from yet another committee of inquiry, the Committee on the Future of Tertiary Education in Australia, an Australian Research Grants Scheme (ARGS) was launched in 1966 to provide peer-adjudicated research grants to academic staff in universities. This committee, chaired by Sir Leslie Martin, recommended the establishment of a national research foundation which would prepare assessments of major research projects for which Commonwealth funds were sought. The government rejected this recommendation but appointed a temporary Australian Research Grants Committee (ARGC) to administer the research grants scheme. The research funds were provided initially on a 1:1 basis by the Commonwealth and State Governments and the arrangement with the ARGC relied on this federal-state co-operation. As the result of a dispute with the state governments concerning the size of their contributions, the co-operative arrangement was ended in 1967 and the ARGC became an advisory body to the Commonwealth Government alone. It is now located administratively within the Ministry of Science and Technology and though not a statutory body, it advises the Minister directly on the allocation of funds to the various applicants. In 1984 $22 million will be distributed. In 1980 the Commonwealth Government announced that it was going to establish a number of 'centres of excellence' in selected universities on a competitive basis and established still another external committee to evaluate proposals. Ten centres have now been established, researching into microelectronics, nerve-muscle interaction, clinical immunology, neurobiology, plant cell biology, cancer and transplantation, environmental fluid dynamics, gene technology, policy studies and mathematical analysis.

There is now in Australia an extraordinary situation in which the universities receive funds for R & D from the Tertiary Education Commission (Ministry of Education and Youth Affairs), from the ARGS (Ministry of Science and Technology) and from the independent committee appointed to consider 'centres of excellence' proposals. Postgraduate research students receive support through the Commonwealth Postgraduate Awards Scheme operated through the Department of Education and Youth Affairs.

The position of the universities compared with CSIRO has, therefore, improved dramatically over the past fifteen years. It is widely held, especially in CSIRO, that the imbalance between these two sectors of research performance has been corrected, and that the undesirable friction which existed between them has been reduced. There is no doubt that it has been reduced, but it has not been eradicated. In its examination of the Australian R & D system, the OECD examiners reported that they had received complaints of insufficient funding 'practically everywhere' in the universities. Emphasizing the fact that the total resources at the disposal of CSIRO for the conduct of scientific research was not much less than that of all research funds of the universities taken together, the examiners concluded that the imbalance between the two was still too great. 'Some of the criticisms,' they suggested, 'and it must be admitted that there were many, arise from envy of the substantial resources which CSIRO enjoys.'[35]

The failure of CSIRO and the universities to co-operate more closely is a recurrent theme in the various reports on the Australian R & D system that have become the favoured mechanism for change in recent years. In 1966 the CSIRO's Advisory Council convened a committee of senior CSIRO officers, academics and Advisory Council members 'to examine co-operation between CSIRO and the universities and to make suitable recommendations.'[36] Apart from recommending informed exchange between CSIRO personnel and university research workers, this committee proposed the establishment of joint CSIRO-university research units, siting of CSIRO laboratories close to universities, sharing of instruments and co-operation in research.

Exactly ten years later, the government set up a Committee of Inquiry into the CSIRO to examine, report and make recommendations on:

- the objectives of the organization and the relevance to the present and future requirements of Australia of its functions as set out in the Science and Industry Research Act (1949);
- the extent to which the current research program objectives and the emphases given them accord with the objectives recommended for the organization;
- existing arrangements and procedures for meeting recommended objectives and discharging recommended functions, with particular emphasis on:
  (i)   the size and diversity of the organization, its organizational and management structure, policies for the employment of staff, and the role of consultative and advisory machinery;
  (ii) the relationship of the organization with government agencies, industry, tertiary institutions, research institutes, and with users of research results;
  (iii) the methods for selecting, reviewing, reporting on, and re-ordering research programs, including the effect of the differing sources of funds;
  (iv) the assessment of results achieved in the light of resources employed;
  (v)   the processes involved in the implementation of research results;
  (vi) the role of the organization in Australia's international scientific relationships;
- the extent to which and the means by which the programs of the organization could attract revenue both to support the conduct of ongoing or intended research and also in return for results achieved in research.

As a result of the recommendations made in the report of this Committee of Inquiry, the Science and Industry Act was amended again. The main function of the CSIRO now is to carry out scientific research for the purpose of

- assisting Australian industry;
- furthering the interests of the Australian community;
- contributing to the achievement of Australian national objectives or the performance of the national and international responsibilities of the Commonwealth;
- any other purpose determined by the minister.

The new Act 'legitimized' CSIRO's involvement with pure research, an activity that it had previously engaged in under the subsection that allowed it to initiate and carry out research in connection with any other matter referred to it by the minister. In his tabling statement, the then Prime Minister said that CSIRO would have a major role in helping Australia meet its international obligations by 'continuing to undertake basic research, such as in astronomy, atmospheric physics and oceanography, and to increase man's knowledge of the region'.

With regard to CSIRO-university links, the Committee of Inquiry proposed the establishment of a joint committee of CSIRO and the Australian Vice-Chancellors' Committee to investigate means of colla-boration; the creation of joint 'centres of excellence'; secondment of personnel between CSIRO and the universities; sharing of equipment; more contracting by CSIRO of research work to the universities and the siting of new CSIRO laboratories adjacent to tertiary institutions. Although the joint Vice-Chancellors/CSIRO Committee was formed in 1978, its activity is judged by ASTEC to be at a very early stage of development.[37] As far as the other recommendations are concerned, none appear to have made much headway. This is not to say that colla-boration between CSIRO and the universities does not take place; it simply is not as extensive as the report would consider desirable.

CSIRO AND SECONDARY INDUSTRY

The Act which incorporated CSIR in 1926 empowered the Council to initiate and carry out scientific research in connection with, or for the promotion of, primary and secondary industries in the Common-wealth. This function remained unaltered in the Science and Industry Research Act of 1949, but was modified as we shall see in the existing legislation which came into force in 1978.

In the early days of CSIR's existence, agricultural problems de-manded immediate attention, and it scored some significant victories in the fight against pests and diseases that plagued crops and livestock. Research results were quickly transferred to farmers through the ex-tension services of the State Departments of Primary Industry and

Agriculture. But CSIR's interaction, and that of its successor with secondary industry, was not so successful. It was not until 1937 that the CSIR became strongly involved with research relevant to the needs of

Table 6.5

Distribution of CSIRO research effort by sector

Sector	% of total research expenditure	
	1981-82	1982-83
Rural industries	33.6	33.5
Mineral, energy and water resources	16.2	16.5
Manufacturing industries	25.4	26.9
Community interests	24.8	23.1

Sources: CSIRO Annual Report 1981-82, pp. 23-24
CSIRO Annual Report 1982-83, pp. 29-30

Table 6.6

Distribution of CSIRO research effort
in the institutes

CSIRO institutes	% of total research expenditure	
	1981-82	1982-83
Institute of Animal and Food Sciences	18.6	19.8
Institute of Biological Resources	26.7	28.3
Institute of Earth Resources	19.6	15.3
Institute of Industrial Technology	16.8	15.1
Institute of Physical Sciences	18.3	21.5

Sources: CSIRO Annual Report 1981-82, p. 25
CSIRO Annual Report 1982-83, p. 31

the manufacturing sector. The delay may have been due to the attitude of its first Chairman, Sir George Julius, who stated

> It is abundantly clear that this country cannot hope to find anything other than the domestic market for the products of its manufacturing industries, except in a very small way. The smallness of the market must preclude for long to come the development of such mass-produced methods as would enable us to reduce our high cost of production, as to be able successfully to compete in the world markets for the supply of manufactured products. The long distance transport must for all time impose a strong handicap upon us in this connection.[38]

Julius argued that since the great primary industries were the very life-blood of the country it was vital that the major focus of CSIR's attention should be in this area. Esplin has dubbed this the agricultural imperative;[39] it persisted until 1937 and, indeed, CSIRO's R & D effort is still biased in favour of the primary industries, as the figures in Tables 6.5 and 6.6 show. When CSIR eventually began to work in areas relevant to the secondary manufacturing sector there was no extension service available to ensure that research results would be transferred and there seemed to be a fear within the CSIR that it might be expected to perform this function. Sir David Rivett, Julius' successor as Chairman, had this to say:

> Am I unreasonable in expressing a fear that to most people the justification for our various national laboratories is to be found in the belief that science is to be the ever-ready handmaid to industry, standing by to help when dividends fall, or a machine fails to function, or a greater yield is wanted, or a better finish? Rare and precious as handmaids are nowadays, I venture to hope that this rather restricted view of the functions of the scientific research institutions will not persist. The *raison d'être* of such laboratories is much more than that. What we need to develop amongst ourselves is the faith that knowledge is worth seeking and worth getting at even though any immediate connection between it and industrial profit may be completely invisible.[40]

Rivett was convinced that the Chairman should be a part-timer, some-one who was not wholly dependent on the government for a living, so that as a last resort he might resign on a matter of principle — to thwart those who would turn CSIR into a handmaiden of industry. Dis-illusioned with the legislation which changed CSIR to CSIRO Rivett himself resigned in 1951. He need not have worried too much about the possibility of CSIRO becoming the handmaid of industry in the post-war years, however. Australian industrialists preferred to buy tech-nology from overseas rather than embark upon risky new technological ventures, and foreign-owned subsidiaries imported the technology of their parent companies. Official government policy of encouraging import substitution by the erection of high tariff barriers and other types of protection maintained the relative lack of interest by Australian industry in developing its own research potential or using the research results of organizations such as CSIRO. In 1964 a member of the CSIRO Executive described his experiences in dealing with Australian industry as depressing and his pleas for more scientific personnel in industry were ignored.[41]

The most recent inquiry into the CSIRO received a number of submissions from industry which commented on the gap between CSIRO and industry, for example:

> It is apparent that the interface between industry and CSIRO's divisions is generally inadequate. The organization's service to Australian secondary industry has fallen short of what is needed, and the links between many CSIRO divisions and industry continue to be tenuous.[42]

The Committee of Inquiry, while accepting that a change in attitude towards manufacturing industry by CSIRO was needed, came to the conclusion that manufacturing industry itself was largely to blame for the poor contact that exists generally between it and CSIRO. The revised Act added a new requirement to the functions of the organization, however—that it encourage or facilitate the application or utilization of the results of its research. Amongst its new powers is the ability to join in the formation of a partnership or company for the purpose of the commercial development of a discovery, invention or improvement of the property of the organization.[43] CSIRO has now been reorganized into five institutes, each under the control of a director, containing a number of divisions and units. The Institute of Industrial Technology conducts R & D aimed at increasing the efficiency, competitiveness and scope of Australian secondary and tertiary industries. In 1979 a Manufacturing Industry Committee was set up to

- assess the scope of CSIRO programs and practices that are of assistance or potential assistance to manufacturing industry;
- encourage institutes and divisions to make new proposals in this field;
- to make recommendations to the Executive on an extended manufacturing industry program; and
- to consider whether any part of such a program is more appropriately undertaken by external contracts.

As a result of its report, a new Division of Manufacturing Technology was created within the new Institute of Industrial Technology to work closely with firms engaged in the metal-processing industries, which account for about 45 per cent of Australia's manufacturing industry. Other initiatives taken by CSIRO in support of Australia's manufacturing capability include the establishment of a research program aimed at the design of very large scale integrated circuits, and the development and marketing of monoclonal antibodies for the diagnosis of human and animal diseases in collaboration with private firms. Though the organization emphasises that the resources it commits to research of relevance to the manufacturing sector will continue to be longer-term than is normally conducted by individual firms, an awareness of the need to ensure transfer of research results to that sector seems to be emerging. Specific measures that have been approved by the Executive include

- strengthening of CSIRO's techno-commercial skills as an aid to the selection of projects;
- direct participation by CSIRO in the further development of its research results, particularly by ensuring that know-how and expertise are made fully available to companies, especially small companies, when developing the research results;
- utilization of patents and licences in such a way as to ensure maximum use of, and benefits from, CSIRO's research results;
- facilitation of mobility of staff both internally and externally;
- development of promotion criteria that incorporate acknowledgement of the importance of practical achievement;
- further promotion and support of research associations.[44]

Importantly, a Planning and Evaluation Unit, headed by a senior scientist, has been established in response to a recommendation made by the independent inquiry into CSIRO, to assist the Executive in the development of strategies, in priority setting and in the allocation of resources. The functions of this Unit are twofold; to provide advice to the Executive, based on analyses of scientific, economic and social data from within the organization and outside it, which will assist the Executive in the discharge of its strategic planning responsibilities, and to advise the Executive, Institute Directors and Chiefs of Divisions on planning, review and evaluation methodology. As part of a process of developing more detailed policies and priorities for its manufacturing industry research, the unit is undertaking analyses of the sector, taking over the work of the Manufacturing Industries Committee. Importantly, the appearance of this high-level policy analysis unit has strengthened, and will continue to strengthen, the CSIRO Executive's capability in this area.

Co-ordination

From a situation of relative *laissez-faire* and pluralism in the management of its scientific affairs Australia has now moved progressively towards greater co-ordination of its R & D effort. Various ministries carry out and externally support R & D and, for a short time a ministerial committee, chaired by the Prime Minister or his nominee, acted as an official co-ordinating body. Since that committee was disbanded in 1976 there has been no co-ordinating committee of ministers or senior public servants; in this respect Australia still lacks a prime element of the science policy machinery that has been described here as the co-ordination system. It does, however, have a high level independent committee that advises the Prime Minister on science and technology matters and this has had a good record of having its advice accepted and acted upon. If any co-ordination of the government's R & D effort takes place it is through ASTEC's role in the budgetary process.

There is a Ministry of Science and Technology with operational and advisory functions but as yet this has no overall co-ordinating role. Indeed, as long as it continues to have operational responsibilities it may not be able to do this since other ministries might legitimately question the right of one line ministry to co-ordinate the activities of other line ministries.

It has to be said, however, that when major reviews of the R & D system take place, the government tends to rely on advisers from outside the official science policy advisory system. It has been pointed out that when the science task force of the Royal Commission on Australian Government Administration was engaged in its review of the entire governmental R & D system it did not consult the interim ASTEC and that body had no say in the selection of the task force members. Subsequently, the CSIRO, the Atomic Energy Commission, the Defence Science and Technology Organizations, the Bureau of Meteorology and the Commonwealth Laboratories have all been subjected to investigation and review by advisory committees outside the official science advisory system. The recommendations of these independent inquirers are usually considered by interdepartmental committees composed of relatively senior public servants and it is at this point, in general, that the official science advisory bodies can influence the final recommendations that go to the Cabinet.

There is no formal mechanism for ensuring that the R & D efforts of the ministries and statutory corporations reflect national social, economic or political objectives. Priorities are identified internally and the R & D program proposals that result are subjected to scrutiny by ASTEC in the budget cycle, where adjustments may be made at the margins. ASTEC is expected to advise on priorities and on the balance of Australian efforts in science and technology. One of the first tasks it undertook was to conduct an overview of science and technology in Australia and this was expected to provide the background from which discussion of priorities for future activities could take place. Following surveys of the practices of certain other countries an attempt was made to identify priorities according to a technique developed by UNESCO.[45] It was concluded that the research priorities obtained have to be interpreted with caution and the results used only in association with additional inputs. In line with some of the comments that have been made in Chapter 3 concerning analytical methods in science policy, ASTEC reported that the exercise served to alert its members to some of the difficulties inherent in seeking to be objective about matters involving complex judgments.[46]

CHAPTER 7

Science policy in the Federal Republic of Germany, Canada and Japan

The Federal Republic of Germany

The government R & D system in the Federal Republic of Germany (FRG) is classified in official literature as pluralist. Scientific and technological activities are conducted in or funded by many ministries and there is a 'high degree of self-administration in science as well as the autonomous studying of programs and establishment of priorities.'[1] In Table 7.1 the 1981 R & D budgets of the top four spenders are shown and it will be seen from this that the dominant force in the FRG's research system is the Ministry for Research and Technology (BMFT), which provides financial resources to the Max Planck Society (MPG), the Fraunhofer Society (FhG), the so-called National Research Centres, the universities and industry. Despite a stated policy of preserving the plurality of research funding sources, an element of co-ordination has gradually been introduced in recent years. At the departmental level, each ministry now has a research co-ordinator, and these draw up statements of the ministries' R & D activities and meet as an interministerial committee under the chairmanship of the Minister of Research and Technology. The Federal Government's research establishments, and institutions like the National Research Centres which are mainly supported by it, whose requirements for federal grants towards their running costs or allocations from the Federal Government for research support exceed DM 5 million, are required to draw up program budgets or other comparable surveys. There is no science budget as such although these departmental program budgets may be regarded as a first step in establishing the conditions under which a science budget might be developed.

The important Ministry of Research and Technology, with its developing powers of co-ordination, has a substantial and increasing ability to direct R & D policy in the FRG through its funding policies,

Table 7.1
R & D budgets of some Federal Ministries (1981)

Ministry	DM million
Research and Technology (BMFT)	5999
Defence (BMVtg)	1563
Economics (BMWi)	1332
Education and Science (BMBW)	863

Source: Factenbericht 1981 Zum Bundesbericht Forschung (Bonn, 1982).

and there are advisory bodies on science policy to give independent advice on the overall development of the national R & D effort. Only the autonomous research institutions stand outside this co-ordination model; as far as these are concerned the Federal and Länder (State) Governments must be content with membership of governing bodies and limited ability to exercise control through the budgetary process. This situation may change, for as the 1978 report of the Federal Government on research states

> In the long run the research establishments will have to be interlinked by means of new co-ordination and co-operation methods enabling them to deal jointly with problems surpassing the barriers of special disciplines and specialized institutions.[2]

The first steps in this direction have been taken and will be discussed under the headings of the two main autonomous institutions. They include co-operation between university and non-university research establishments in 'priority' research areas, co-operation between universities and the MPG in new project groups, and co-operation between the National Research Centres. Thus, we may say that although the West German research system is classified as pluralist there is a greater concern for co-ordination than one would normally expect to find in such a system.

THE MINISTRY OF RESEARCH AND TECHNOLOGY (BMFT)

The aims of the FRG's R & D effort are clearly stated in the government's *Report on Research*, an impressive policy document published by the Minister for Research and Technology every two years. They are

- to extend and deepen the level of scientific knowledge;
- to maintain and increase the efficiency and competitiveness of the economy;
- to conserve resources and preserve the natural requirements for life;
- to improve man's working conditions and the well-being of civilization;

- to recognize the implications and correlations of technological developments, to discuss and balance their opportunities against risks,
and to substantiate decisions on the utilization of technologies.

In the pursuit of these aims the Federal Government spent DM 10 897
million in its ministries on R & D in 1981. The Ministry of Research
and Technology (BMFT) accounted for DM 5999 million (55 per cent).

The BMFT had its origins in the Federal Ministry of Atomic Affairs,
established in 1955 to promote research into, and the use of, nuclear
energy for peaceful purposes. One of the main reasons for the establishment of this ministry and for the subsequent addition of space
research and other high-technology activities to the portfolio was the
intention to close the 'technology gap' that existed between Germany
and other industrial nations as a result of the Second World War. The
ministry became the Ministry for Scientific Research in 1962. In 1969
the Federal Government, on acquiring new responsibilities in the
higher educational and educational planning sectors, created a new
ministry — the Ministry for Education and Science, which incorporated
the Ministry of Scientific Research. Finally, in 1972, the Education
and Science branches were separated and the Bundesministerium für
Forschung und Technologie came into being.

The BMFT has a total staff of about 600 people and is organized
into divisions and subdivisions. Its responsibilities, as laid out in the
administrative orders, are simple. The ministry promotes research
activities and new technologies which contribute decisively to shaping
the future of West German society, and directs its activities towards the
government's aims listed earlier. Of these aims, the one that refers to
the efficiency and competitiveness of the economy is the most important, as the breakdown of the 1978 and 1979 budgets show (Figure
7.1).

In implementing its terms of reference, the BMFT has three
principal responsibilities. First, it promotes individual projects;
second, it provides basic financing for the National Research Centres
and the autonomous institutions; and third, it is responsible for the
management of the FRG's international scientific obligations.

The BMFT promotes individual projects in both higher education
and in private enterprise. Allocations to higher education, amounting
to about 10 per cent of the ministry's available funds, are made in high
cost areas such as high-energy physics or astrophysics or where research
programs that need access to observatories, accelerators and nuclear
reactors are mounted by universities in co-operation with the National
Research Centres. The bulk of the BMFT's project promotion is in the
industrial sector. However, private enterprise is the largest single
contributor to R & D funding in the FRG. As shown in Table 7.2, of
the DM 30 billion spent on R & D in 1978, more than DM 15 billion

Figure 7.1

Aims and priorities of the Federal Government's research and technology policy

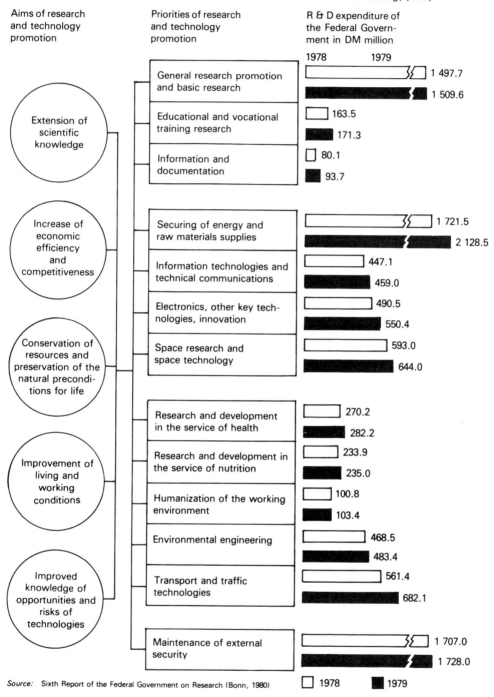

Source: Sixth Report of the Federal Government on Research (Bonn, 1980)

were accounted for by the private enterprise sector. Industrial R & D
expenditure is concentrated in a number of key industries. Chemicals,
steel construction/mechanical engineering, precision engineering,
vehicle building, electrical engineering and optical industries account
for about 90 per cent of all industrial R & D, industries which are
generally the most export-intensive and have the largest firms. Only
about 5 per cent of industrial R & D is carried out by firms with less
than 1000 staff.

Table 7.2
Total R & D budget in the FRG for 1978 by funder and performer (DM million)

Funder Performer	Federal Government	Länder	Industry	Others
Industry	3600	50	15 096	520
Universities	935	4250	100	—
Non-university research institutions	4010	1340	140	360
Total	8545	5640	15 336	880

Up to about twenty years ago, government aid for innovation in the
private sector was almost non-existent. By 1977 the situation had
altered dramatically. In that year the Ministry for Defence (BMVtg)
provided DM 2930 million in aid; the BMFT provided DM 1400
million and the other ministries, including the Ministry for Economics
(BMWi) provided DM 1600 million. The share of R & D expenditure
by the industrial firms themselves has declined over this period. In
1962 this was more than 85 per cent; it is now less than 60 per cent. The
government has advanced many reasons for this rapid increase in federal
promotion of R & D. Modern key technologies — data processing, electronics, materials engineering — are increasing in importance and West
German industry must keep pace with these developments if it is not to
lose market share. The level of scientific and technical know-how is
increasing rapidly in the civil as well as the defence sectors. This know-
how must be continually mastered. The scope and cost of R & D are
increasing faster than the capacity of industrial enterprises to absorb
them. The short-sighted concentration of managements on immediate
returns on investment in R & D has to be overcome. The general public
and the government are becoming more and more aware of the need to
secure supply of energy and raw materials, to protect the environment,
to improve public services in transport, traffic management, health
care, urban planning, housing information and communications.
Government development contracts for military goods and its spin-off
into the civil sector in one country have a bearing on the competitive

situation and bring about counterbalancing promotional activities in other countries. A defensive R & D posture must therefore be maintained.

The fifth report of the Federal Government on research outlined the principles governing the promotion of industrial R & D. Government support is granted when private enterprise is not strong enough to initiate the necessary measures or to initiate them in due time. This applies especially whenever

- the scientific and technical, as well as the economic, risks have to be rated as very high;
- large funds are required;
- developments will take a very long time so that profits cannot be expected in the foreseeable future;
- the market underrates more advanced technological solutions since it responds not so much to the future as to the prevailing conditions as far as demand, shortage and constraints are concerned; or demand does not suffice to bring about new technological solutions which are exclusively or predominantly in the interest of the general public or are applied in the public sector.[3]

It will be seen that public support for industrial R & D is justified in terms of the market failure theory. In following the guidelines laid down by the Federal Government for the promotion of industrial R & D, the BMFT spends most of its substantial funds on applied research in industry. Preference has been given to the technique of promoting concrete projects and programs (direct promotion) instead of offering funds on a broad uniform basis for research costs incurred by enterprises (expenditure-related promotion). Key technologies in the electronics, communications and energy-related industries receive special support, and direct promotion allows this relatively easily. Research and development projects are promoted with grants amounting as a rule to half of the costs involved if they comply with the goals and conditions of BMFT's major programs, which are drawn up and published by the BMFT. One example is the Energy Research and Energy Technologies Program, which sets out for potential contractors the basic principles and objectives of the program together with an analysis of the current situation. Within the major program there are various sub-programs, for example, coal, nuclear, solar, new sources of energy, and these in turn are broken down into projects. The programs, sub-programs and projects are all given provisional four-year budgets.

By promoting industrial R & D projects the Federal Government acquires the right to use all of the results obtained from such activities free of charge for its own purposes, including the right to patent them.

The enterprises engaged in such sponsored activities are, in certain cases, obliged to grant licences to third parties against an appropriate royalty, and if there are unusually high receipts from the licensing and from innovations arising out of the results, the government is entitled to a share. In the case of R & D projects oriented to the market, the enterprise receiving government assistance is obliged to repay the amount it received should the project prove to be commercially successful.

In addition to its promotional activities in the area of industrial R & D, the BMFT provides basic financing for the twelve National Research Centres, and shares responsibility with the Länder and others for financing the Max Planck Society and the Fraunhofer Society. The National Research Centres are concerned with nuclear research and technology, biology, medicine and environmental research, aerospace research, mathematics and data processing, cancer research and biotechnology. They employ about 16 000 people, 4000 of whom are research scientists and they have a total annual budget that is now more than DM 2 billion. They carry out long-term R & D, especially in those fields in which BMFT major programs have been published, and in priority areas of 'big science'. Each Centre has a board of directors, chaired by a BMFT representative, which convenes twice a year to discuss progress and to approve forward budgets.

Priority identification in the BMFT
Project promotion is the most appropriate of the three major areas of BMFT activity to discuss priority setting. Research policy is considered within BMFT to operate at five levels:

- Overall research policy is executed by the minister.
- Research policy as it affects the establishment of major programs is the responsibility of divisional heads. Currently, fourteen major programs are in operation within BMFT.
- Sub-programs are the responsibility of deputy heads of divisions. 125 are currently in operation.
- Activities are defined as particular approaches to sub-programs, for example, 'direct conversion' in the solar energy sub-program of the major energy program. At present, there are 360 activities in progress.
- Projects within these activities, of which there are 6500, are, together with activities, the responsibility of project managers.

For some years now the BMFT has been adopting a procedure of 'project management' and 'project accompanying organizations' for the implementation of research programs that come within its scope. The 'project managers' are institutions responsible for the preparation and execution of a particular decision on promotion through the planning and supervision of project schedules, the co-ordination of various individual projects and the evaluation of results. The National Research Centres and, in some cases, industrial research associations function as project managers. At present there are fourteen of these acting on behalf of the BMFT in the supervision of 2500 projects. 'Project accompanying organizations' are research institutes, industrial firms or individuals who, in one way or another, concern themselves with the implementation of the project and advise the ministry on its progress and evaluation of its results. In all, about 1300 advisers are used by the BMFT in the management of its major programs. They work through ad hoc committees, panels of experts and project committees and provide the ministry with expert advice on project applications and on the execution and guidance of major projects.

Priorities may be influenced at a number of levels. When there is a change of minister, the incoming minister may bring with him certain ideas that the BMFT will implement. For example, a program in the humanization of the work environment was commenced on the initiative of Hans Matthofer, Minister from 1974 to 1978. Volker Hauff, his successor, initiated a plan to slow down the ministry's program for the support of the computer industry and established a program dealing with the implications of and problems caused by modern information technology. A second opportunity to change priorities occurs when programs are being reviewed at the end of each four year period, when a searching analysis of the progress made and future prospects is carried out. A third opportunity is presented during the annual budgetary procedure in which the Minister's budget for the fourteen program areas is presented to Parliament for approval, program by program and project by project. During the in-house deliberations with the Minister on the budget, priorities may be changed before these proposals are presented to Parliament. Finally, priorities may be changed as a result of the feedback coming into the BMFT from advisers who are evaluating new applications.

The BMFT relies, therefore, on expert advice coming from internal and external sources for the determination of priorities. This advice will often be influenced by the socio-economic climate and by government policy. Formal methods of priority assessment are not used in the determination of areas that will receive priority. Only one attempt was made, in the 1970s, to apply such techniques — to decide whether to fund projects in high-energy physics, radio astronomy and solar physics. A committee comprising scientists, journalists and sociologists was formed and a modified Delphi technique was used. The result was disappointing and certainly no better than that which might have been expected from informed discussion with the scientists involved in the proposals and with ministry advisers. The ministry now prefers to rely on these advisers, on discussions with scientists involved with proposals, on in-house discussion and on the results of assessments, sometimes economic, that it commissions from professional institutes.

THE MAX PLANCK AND THE FRAUNHOFER SOCIETIES

The Max Planck Society (MPG) is an autonomous research organization almost entirely devoted to basic research. It has 49 research institutes, 40 of which are in the natural sciences and differing widely in size and objectives. In all, 10 000 people are employed in MPG institutes, of whom about 4000 are scientists. The main concern of the MPG is to further those fields of research which, because of their interdisciplinary character, cannot yet be adequately studied at universities or which, because of their special requirements for staff or equipment, do not readily lend themselves to university research. In particular, the functions of the MPG are

- to support developing research areas, especially marginal areas that are only slowly gaining access to the universities where work is subject to teaching demands;
- to develop new types of institutes and to further those research projects that are so extensive and complex that universities hesitate to undertake them;
- to relieve eminent people from teaching duties so as to enable them to make full use of working facilities designed to their requirements;
- to provide teaching facilities for young people by grants and scholarships.

The Federal and Länder Governments provide about 85 per cent of MPG operating and capital expenditure, usually in a 50/50 ratio. In certain areas such as astrophysics, observational astronomy, plasma physics and solid state physics, the federal contribution rises to about 90 per cent.

Important decisions concerning the MPG's overall planning are taken by its Senate, on which the Federal and Länder Governments are represented. The MPG institutes co-operate with the universities in a number of ways. The Chairman of the Conference of University Rectors and the President of the German Research Society can attend MPG Senate meetings as observers, university professors are eligible for election to the Senate and MPG scientists can acquire university doctoral degrees or have honorary professorial status. In addition, the MPG provides places in its institutes for a large number of university students who are completing their doctorates, and MPG scientists sometimes accept teaching duties as associate lecturers in order to introduce new research ideas and to co-operate in research programs in special areas.

For many years MPG research programs were established in particular areas simply because opportunities arose to appoint eminent scientists. In 1972, however, the MPG Senate Committee for Research Policy and Planning was established and charged with the continual collection and evaluation of information on current and projected research programs. On the basis of this information, proposals are now put forward for the identification of priorities, the determination of the size of research projects to be started and the founding and abolition of institutes. Decisions on the establishment or closure of research areas are made in close consultation with other major organizations—the Science Council and the German Research Society, for example, and with the Federal and Länder Governments. In order to achieve flexibility of research programs, departmental structures have been introduced, directors' terms of office have been limited to seven years and visiting committees of scientists with international reputations have been established to advise the institutes in their research and to give regular reports on their work to the President of the MPG. These reports are essential to the work of the Senate Research Policy and Planning Committee.

The Fraunhofer Society (FhG) is a mission-oriented organization employing about 1700 staff which makes an extensive research capacity available to industry and the various governments to assist technological development and also increases the technological capabilities of small and medium sized firms by providing an R & D contract service. Contract research with FhG institutes gives the contractor exclusive access to the results. The contracts are carried out at cost and small and medium sized firms can resort to a special public fund for part of the contract fee. The Society is enabled by this fund to advise firms on the application of new technologies and the market prospects for new products. On the matter of priorities within the FhG, the fifth report of the Federal Government on research had this to say:

> As far as the Fraunhofer Society is concerned the efficiency control is to be guaranteed by the financing scheme because in this case the amount of public grant is made dependent on the amount of contractual research carried out, which is thus assigned the function of an indicator of success.[4]

The FhG intends in future to step up its efforts to provide contractual research facilities for small— and medium—sized firms and help such companies to pinpoint technical problems and solve them through science and technology. It will also continue to foster the transfer of know-how acquired by defence contractors to the civil sector.

THE GERMAN RESEARCH SOCIETY (DFG)

The universities in the FRG receive their recurrent funds from the Länder Governments and their research funds from the Federal

Government through the German Research Society (DFG), the BMFT and other ministries. The DFG is an autonomous granting body which is supported by both the Federal and Länder Governments and is administratively located within the Federal Ministry of Education and Science. It was founded in 1951 to

- provide financial assistance for research projects;
- encourage co-operation among research workers;
- advise the government and public authorities on scientific matters;
- encourage close relations between scientific and industrial circles;
- promote German co-operation in international scientific life;
- give encouragement to young scientists.

The Society is organized into a General Assembly, which takes decisions on scientific policy, a Senate, which co-ordinates scientific activities and gives advice to the Federal Government, a Board of Trustees and a General Committee, which takes final decisions on research grants. The Federal and Länder Governments are represented on all DFG bodies that take financial decisions.

According to its terms of reference the DFG promotes research in all fields and disciplines, and it uses three basic procedures in carrying out its task. Under the normal procedure research grants are awarded to individual scientists or groups of scientists as a result of their unsolicited proposals judged by peer review. In its priority programs the DFG selects projects on the basis of social and economic need as well as the criteria of scientific merit and the need for co-ordinated efforts to reach certain goals. Priority programs are suggested by scientists, either individually or through their organizations and co-ordination with the Federal Government's own priority programs is regularly sought. Finally, in the special research programs, support and co-ordination is provided for a substantial group of scientists (usually more than twenty). The initiative comes from individual universities or groups of neighbouring institutions working in a selected field. The permission of the German Science Council and of the appropriate Länder Minister must be sought before a special research area can be established.

The DFG also establishes research units consisting of small groups of scientists, following a recommendation from the Science Council in an effort to close gaps in fields where teamwork, especially interdisciplinary co-operation, is required for up to five years and where research efforts must extend beyond the framework of a priority project. These research units can be seen as crystallization points for the special research areas.

Periodically the DFG issues a major policy document called the Grey

Plan in which its triennial financial projections are recorded. The Grey Plans act as policy documents as well and are generally critical of the financial provisions made in the preceding triennium for the support of undirected research in the FRG. However, in an attempt to co-ordinate the Federal and Länder Governments' priorities with its own, the DFG is now seeking submissions from the governments so that its planning mechanism can take due note of the social policy aspects that governmental priorities usually contain.

CO-ORDINATION

The science policy system of the Federal Republic of Germany is des-cribed in the official literature as pluralist; this is true if pluralism is taken to mean that many ministries have responsibility for intramural performance and extramural funding of R & D. Elements of co-ordination are being introduced, however, and the principal agent of co-ordination is the BMFT. Each ministry with R & D responsibilities has a research co-ordinator and all research co-ordinators meet as an interministerial committee, under the chairmanship of the Minister for Research and Technology, to discuss the program budget that all ministers are now required to prepare if their R & D expenditure derived from the Federal Government exceeds DM 5 million. All departmental R & D projects send reports to a central databank run by the BMFT.

The BMFT is the major source of policy advice to the Federal Government on science and technology. Through its major programs this powerful ministry is able to respond to the priorities given in the report of the Federal Government on research. Although the auto-nomous research institutions (MPG and FhG) and the universities stand outside this co-ordination framework, steps are under way to link these more effectively into a national science policy. In the fifth report of the Federal Government on research a clear indication was given that these institutions would be expected to contribute more directly to problems of pressing national interest.

Canada

Few countries in the world have undertaken as many analyses, evalua-tions and re-evaluations of their R & D systems as Canada has done, and the objective has generally been to improve the country's second-ary industrial trading position. Since the beginning of this century, the main dynamic factors in the Canadian economy have been exploita-tion of natural resources and primary processing of raw materials. The manufacturing industries, relying heavily on imported technology, have merely responded to the needs of an expanding and protected domestic market, and the service sector has been able to expand rapidly because it could respond to rising affluence without being heavily exposed to market forces or international competition. Since 1916, when the newly-formed National Research Council (NRC) was asked to select the most practical and pressing problems indicated by industrial necessities for earliest possible solution, stimulation of the manufacturing sector has been a major Canadian preoccupation.

There is a strong belief in Canada that innovation, economic growth and international competitiveness in the manufacturing sector will be achieved by investment in R & D. Politicians, science policy makers and the scientific community seem to be obsessed with Canada's low position in the international league table of Gross National Expenditure on Research and Development (GNERD) as a percentage of gross national product and with the reluctance of Canadian industrialists to increase investment in R & D. Writing in 1970, the Senate special committee on science policy, the Lamontagne Committee, observed that

> if we look at the international situation, the uniqueness of Canada's position becomes obvious. Canada is at the bottom of the list as far as R & D performance by industry is concerned, but at the top when it comes to government and non-industrial sectors.[5]

Seven years later, Mr Faulkner, then Minister of State for Science and Technology, saw no change. 'The Government is aware that Canada's national research effort is less than half that of other industralized nations and that the distribution of efforts among the three performance sectors is the inverse of most other nations.'[6] The ministry and others continue to press for greater participation by Canadian industry in the funding and performance of R & D, and the Federal Government has set a GNERD target of 1.5 per cent of the GNP by 1985, half of which it anticipates will be funded by private industry. In 1970, private industry funded 31 per cent of Canadian R & D; in 1980, this had risen by a mere 3.5 per cent so, in this one respect at least, Canadian science policy has failed to achieve its aim. And despite many policy statements and exhortations, the GNERD/GNP ratio has stubbornly remained at about 1 per cent up until 1982. The various changes that have taken place in the executive and advisory systems for R & D have been stimulated by the perceived need to alter this situation and there are now strong indications of success.

Until the early 1960s, Canadian science policy was, in the words of the Lamontagne Committee, a science policy by accident. Government involvement in the performance of R & D was, and continues to be, disproportionately high; policies for the stimulation of R & D in the private sector were almost non-existent, and there was an orientation towards pure science even in the Federal Government's establishments.

Since the mid-1960s, however, Canada has been engaged in a rather lengthy process of acquiring a science policy machinery, and this can now be said to be of the concerted action variety. The main elements of the advisory, executive and operational system for Canadian governmental R & D are shown in Figure 7.2.

THE MINISTRY OF STATE FOR SCIENCE AND TECHNOLOGY (MOSST)

In 1960, the Canadian Government established a Royal Commission

Figure 7.2

Main elements of the Canadian R & D system

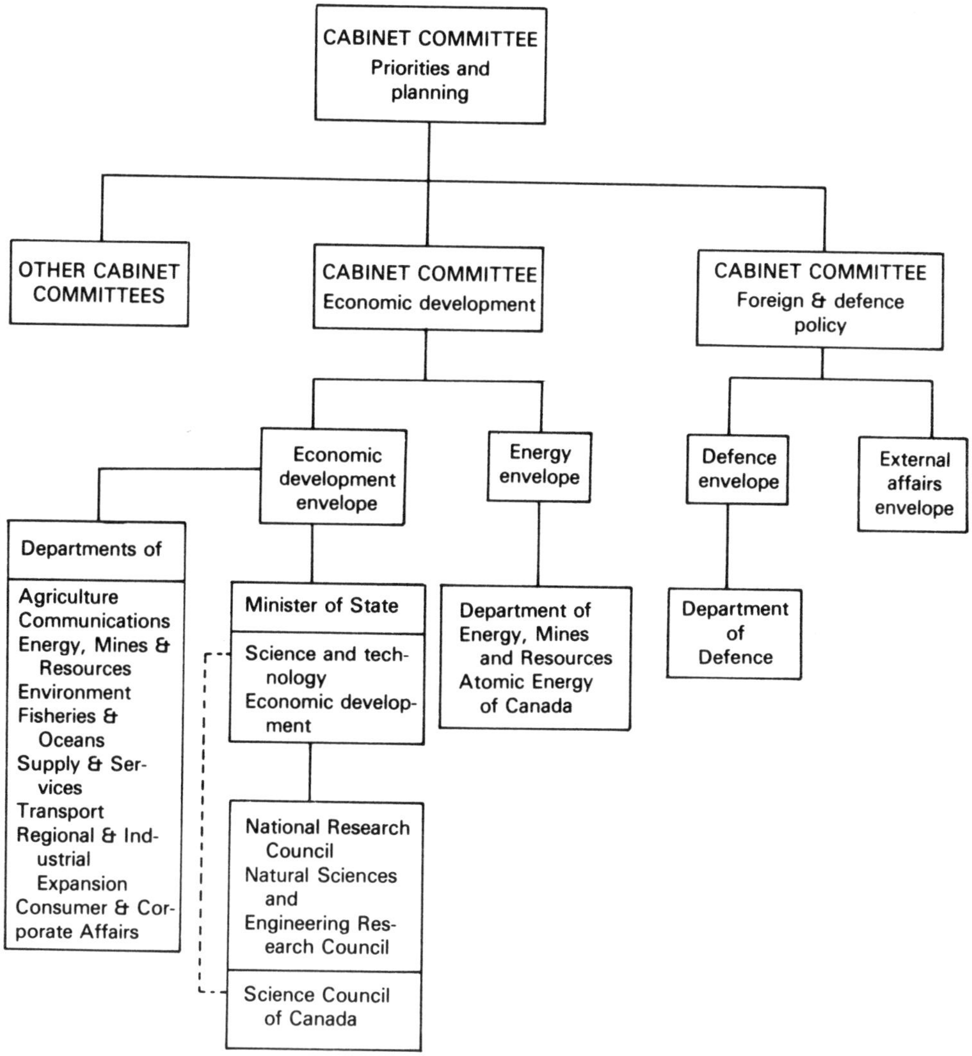

on government organization under the chairmanship of J.G. Glassco, which, as part of its work, undertook an extensive examination of the Federal Government's scientific activities and their organization. The Glassco Report significantly altered the course of science policy in Canada. One of the main recommendations was the creation of a Central Scientific Bureau to act as a science secretariat to the Cabinet, with duties including assembly of information and the conduct of studies in the field of science policy. A second recommendation was for the establishment of a National Scientific Advisory Council to mobilize the independent views and advice of knowledgeable groups both inside and outside the government. The Council would have staff support from the Central Scientific Bureau.

A science secretariat was duly established in the Privy Council Office in 1964 but the passage of legislation to set up the Advisory Council was delayed and did not come into effect as the *Science Council of Canada Act* until 1966. This delay made it necessary for the science secretariat to become a surrogate advisory council, and initiate an appropriate work program. During the first two years of the joint operations of the science secretariat and Science Council, it became clear that the arrangement would not work owing to the confidentiality of the secretariat's operations as adviser to Cabinet and the openness of the Science Council's activities. Eventually the science secretariat was divided, one half remaining in the Privy Council Office and the other half becoming the nucleus of the Science Council's own permanent staff. This arrangement continued until 1971, when an Order-in-Council established the Ministry of State for Science and Technology (MOSST). This filled a 'vacuum at the top' identified in the first report of the Lamontagne Committee, issued the previous year. A Ministry of State can be created for designated purposes by proclamation, and is charged with responsibilities for developing new and comprehensive policies in areas where the development of such policies is of particular urgency. The new ministry was charged with responsibility for developing policies for:

- the most appropriate means by which the government may have a beneficial influence on the application and development of science and technology in Canada;
- the co-ordination of programs and activities involving science and technology with other policies and programs of the government;
- the fostering of co-operative relationships between science and technology and the provinces, public and private organizations and other nations.

In addition, it was to assist departments and agencies of the Government in developing their own science policy contributions, initiate research analysis and policy studies of the impact of science and technology on society, and determine and promote the use of methods of assessing the effectiveness of scientific policies and programs.

During 1972 and 1973 the MOSST kept a fairly low profile, performing its primary function of adviser to Cabinet on policies and programs relevant to science and technology. Despite its broad terms of reference and its responsibility for the overall formulation of policy and co-ordination of the government's science and technology activities, it was generally expected to provide a narrow-based service to departments and agencies, assisting them in the preparation of their scientific programs. It suffered from instability at the top, with frequent changes of Minister and Permanent Secretary, and did not have the policy instrument of a science budget, so the powers of the new ministry were limited. The Lamontagne Committee, in the third volume of its Report, issued in 1973, came to the conclusion that the MOSST was unable to fulfil the larger role demanded by a concerted action system of science policy. As long as it had to rely on the voluntary co-operation of departments and agencies, all of which were naturally jealous of their own prerogatives, the ministry could play no useful co-ordinating role. The senators recommended that an external task force be set up to review the organizational structure of the MOSST and proposed that its powers be strengthened by the establishment of a special budget procedure in which departments and agencies would prepare separate R & D expenditure proposals and submit these to the MOSST for review and assessment before final Treasury Board approval.

The proposed task force was never established but, in 1974, the government approved new terms of reference for MOSST. The ministry would have a stronger role in the formulation of new science-oriented policies; a special budgetary procedure would be developed for examining and approving departmental and agency science expenditure proposals, to be published separately; and it would be given the additional responsibility of reviewing and assessing science expenditure proposals before final approval. Thus, the groundwork for the eventual establishment of a concerted action science policy machinery had been laid. Under these procedures the departments would have to establish a dialogue with the MOSST about their programs well before the stage of formal submission of annual estimates for approval and a science budget, which the MOSST would have the power to review and assess, would be the outcome.

However, in its final report, published in 1977, the Lamontagne Committee was highly critical of the way in which the Federal Government and the MOSST had implemented its recommendations for reform of the budgeting system. Since the beginning of the 1975-76 fiscal year, departments and agencies had been required to prepare separate R & D expenditure proposals and the MOSST, in theory, was authorized to evaluate these. But the reality seemed to be quite different. The secretary of the MOSST claimed in his evidence to the Committee that it received most of the R & D expenditure proposals at the same time as the Treasury Board received them for final approval. Only 25–30 per cent of the estimates were sent directly to the MOSST, which simply did not have the time to review and assess them properly, and even less time to look at the overall picture presented by the government's R & D system so that it could assess this in the light of government objectives. The so-called science budget was hardly being used as an instrument of policy at all, since the MOSST was unable to shape the content and direction of the government's scientific effort to any significant extent.

The final report of the Lamontagne Committee reiterated its views on budgetary procedure and the respective roles that the MOSST and the departments should play in it, but the situation has not been improved to any great extent. The concerted action model of science policy, strongly advocated by the Committee, has not been implemented but recent budgetary reforms may have rendered the introduction of such a system unnecessary in any case. During 1979, the government introduced a Policy and Expenditure Management (PEM) system which, it is claimed, will provide effective interdepartmental co-ordination and ensure that policies can no longer be determined in isolation from expenditure considerations and expenditures cannot be planned without reference to policies and objectives.[7] The system is designed to assist ministers to set and achieve their policy objectives by co-ordinated advance planning. Policy and resource decision-making are integrated and objectives and broad directions are identified several years into the future.

The PEM system has two essential requirements. First, a longer-term Fiscal Plan, encompassing government revenues and expenditures over a multi-year period, is prepared. This sets out the overall resource constraints within which policy choices must be made. Second, specific expenditure limits—the resource envelopes—must be estimated for policy sectors, and these expenditures must be consistent with the Fiscal Plan and with the government's priorities. The system is managed by a number of cabinet committees. The Cabinet Committee on Priorities and Planning establishes the government's general direction and, in the light of anticipated revenues, determines the multi-year Fiscal Plan. This committee's decisions are taken on the recommendations of the Minister of Finance in consultation with the President of the Treasury Board. Four cabinet policy committees are responsible for the development of sectoral strategies for policy and program development within the multi-year resource envelopes established in the Fiscal Plan. These committees receive directions from the Cabinet Committee on Priorities and Planning on policy and program development and they must take into account overall government priorities and objectives. The departments must prepare multi-year and budget-year operational plans and provide strategic overviews to the appropriate cabinet policy committee. Most of the government's science and technology activities are under the jurisdiction of the Cabinet Committee on Economic Development (see Figure 7.2).

The MOSST Strategic Overview for the period 1983/84 — 1985/86 took as its main theme the integration of science and technology policies with other policies, especially the economic development policies of the government.[8] Setting the government's R & D target, the transformation of this target into an R & D planning framework, and the establishment of economic development priorities to which science and technology were closely related were identified as some of the most important steps that had been taken in recent years to achieve Canada's economic objectives. The results of these initiatives were encouraging. The downward trend in R & D expenditure has been arrested and in the vital industrial sector this expenditure has increased rapidly (see Table 7.3).

Table 7.3

GNERD by funder 1979-1981

Funder	1979	1980	1981	Average rate of growth Planned	Actual
		Can. $ million			
Federal Government	936	1105	1254	17%	16%
Provincial Government	173	194	213	19%	11%
Industry	1034	1221	1481	27%	20%
University	344	346	387	9%	6%
Other	144	163	183	9%	13%
	2631	3029	3518		

Source: The Government of Canada's Investment in Science (Ottawa, 1982)

In September 1982, the integration of science and technology policy with economic policy took a significant step forward in Canada when the President of the Treasury Board became Minister of State for Science and Technology and Minister of State for Economic Development. In announcing the appointment, which occurred as part of a reshuffle of ministerial responsibilities, Prime Minister Trudeau said that the 'crucial importance of science and technology to our economic growth make it necessary to ensure there is even greater integration of these policy sectors.'⁹ Subsequently it was announced that Canada's GNERD had reached an all-time high of 1.34 per cent of GNP in 1982 and the preliminary figures for 1983 were indicating a continuation of the upward trend in spending on R & D. The industrial sector was setting the pace with an investment in 1982 of Can$ 2.6 billion, an increase of 28 per cent on the 1981 figure.¹⁰ This result must be seen as a major triumph for a nation that appeared to be losing the battle against underinvestment in industrial R & D.

THE SCIENCE COUNCIL OF CANADA

The Science Council of Canada was established as a Crown Corporation in 1966 and given certain responsibilities in the area of science policy. Under Section 11 of the Council's Act it was charged with

- assessing Canada's scientific and technological resources requirements and potential;
- determining the adequacy of scientific and technological research being carried out and consideration of priorities that should be assigned to specific areas;
- advising on the effective development and utilization of scientific and technological manpower in the country;
- advising on the responsibilities of departments and agencies of the government in relation to those of the universities, private industry and other organizations in furthering science and technology in Canada;
- making recommendations by the publication of reports.

Only months after its establishment the Science Council was asked by the Minister of Mines and Technical Surveys (now Energy, Mines and Resources) to consider a proposal for an intense neutron generator and make recommendations as to whether the project should be approved. The Council recommended in favour of the project but the government subsequently decided not to proceed. Few matters have been referred to the Council by the government since then.

From 1968 to 1971 the science secretariat continued to give confidential advice to the Cabinet, while the Science Council concentrated on producing reports and special studies on science policy problems. When the MOSST was created in 1971 the possibility of any direct influence by the Council on government science policy was lessened considerably. Unlike the MOSST, the Science Council is not part of the Federal Government structure and has no executive or co-ordinating role. The only visible signs of its activities have been reports, complied by Council working parties, on matters such as *Policy Objectives for Basic Research in Canada* or *Towards a National Science Policy for*

Canada, and the special studies that are carried out by Council permanent or seconded staff.

Since the MOSST is charged with the 'initiation of such research analyses and policy studies as may be required', there has always been some overlap in the analytical activities of the two bodies. Legislation known as Bill C-26, which was passed in 1976, attempted to emphasize the partition of the Council and the MOSST by amending the Council's terms of reference to:

- assess the nation's scientific and technological resources, requirements and potentialities;
- increase public awareness of
  (a) scientific and technological problems and opportunities; and
  (b) the interdependence of the public, government, universities and industry in the development and use of science and technology.

Reviewing the first fifteen years of operation in the Annual Review for 1981, the Chairman claimed four roles for the Council. It is a national, non-partisan advisory body, influenced neither by sectoral nor jurisdictional considerations; it does not see itself solely as a federal adviser. The Council has a public role, facilitating dialogue between the decision-makers and the public, bringing the implications of science and technology to public attention and helping interested parties to define their positions and make appropriate representations to government. The Council has an early-warning role, alerting Canadian society to problems and opportunities arising out of developments in science and technology. The Council has adopted an international awareness role which places science policy developments in Canada in an international context. The Science Council values its autonomy and sees the concept of an independent science council, free from governmental constraints and operating at arm's length from government departments and agencies, as a model approach to the development and implementation of science policy. Free from much of the day-to-day pressures of government work and able to express independent opinion, the Council can, it is claimed, concentrate on the development of new policies for the future.[11]

However, the Lamontagne Committee, in the fourth volume of its report, was not entirely happy with the direction that the Science Council had taken. The Committee had been informed by the MOSST as early as 1975 that the government expected the Council to concern itself more with public awareness of science and its implications for society and, while welcoming this move, the Committee pointed out that if the Council was to become a sophisticated public information bureau, Canada would be without an impartial enlightened advisory body to provide guidance to the government on important technological issues. It is not surprising that the Lamontagne Committee took this stance. Its recommendations to the government were couched in terms of the concerted action model of science policy. In this model, as we have seen, the central co-ordinating minister is advised by a powerful external committee. The Science Council has never adopted that role and it is doubtful if it ever could, bearing in mind the circumstances under which it evolved.

Priorities in the Science Council of Canada

The Science Council of Canada has turned its attention to the matter of priority determination on several occasions. In its fourth report, published in 1968, the Council attempted to set out general policies for the effective use of science in Canada. Starting with the axiom that the value of any scientific enterprise to a society is determined by the social, cultural and economic goals that the society seeks, a frame of reference was built up in four stages. These were:

- identification of a set of goals which, while not comprehensive, appeared to contain the main aspirations of most Canadians;
- identification of the various factors on which the ultimate attainment of each goal depended;
- identification of the contributions that science and technology could make towards the attainment of the goals;
- identification of the conditions that would permit these contributions to be made.

Six goals were chosen to provide the focus for policy discussions. These were:

- national prosperity;
- physical and mental health and high life expectancy;
- a high and rising standard of education readily available to all;
- personal freedom, justice and security for all in a united Canada;
- increasing availability of leisure and enhancement of the opportunities for personal development;
- world peace based on a fair distribution of the world's existing and potential wealth.

The Council did not elaborate on the mechanisms for making policy decisions on the basis of these goals, nor is it evident that it used the identified goals in the advice contained in the rest of its fourth report. All we are told is that the maintenance of a prudent balance of the resources assigned to the various goals will be important and that the Science Council will be considering this problem of resource allocation on a continuing basis. The main recommendation arising out of this report was that most new undertakings in Canadian science should be organized as large, multidisciplinary mission-oriented projects having as a goal the solution of some important economic or social problem and in which all sectors of the scientific community participate on an equal footing. In deciding on the establishment of a major program the following criteria would have to be met:

- the objective selected for each major program must be of real importance to Canada, and perhaps even peculiar to Canada;
- no major program should duplicate work already under way in other developed nations;
- there must be some demonstrable prospect of direct social or economic benefit which in an overall view would be commensurate with the resources invested;
- the scientific and technological challenges must be fundamental and far-reaching enough that they will not be quickly exhausted, and yet in general not so abstruse that there is little hope of tangible progress within ten or twenty years;
- the unpredictable quality of research and the openendedness of the future must be clearly recognized;
- projects must be mounted on a sufficiently large financial scale that the various R & D groups formed to attack the special problems will be of above-critical or viable size, and will have reasonable prospects of a steady diet of challenging projects within their range of competence;
- the choice of a program should be based on a conjunction of need and of scientific or technological opportunity.

The Council recommended that a co-ordinating body be created for each major program, envisaging that substantial funds would be made available. The major programs never eventuated in the form proposed by the Science Council but there is little doubt that the special programs referred to later have their origins in the concept. In its fourth report, the Science Council recommended that two prototype major programs be established in space research and in water resources, and that others be planned in transportation, urban development, computer applications, energy, oceanography and marine science and integrated resource management. The five special programs now in force under the direction of interdepartmental committees are contained in that list in one form or another.

In 1972 the Science Council issued a report entitled *Policy Objectives for Basic Research in Canada* in which criteria for the selection of projects in basic science were laid out. These criteria drew heavily upon those developed by Weinberg. No guidelines were given by the

Science Council on the way in which the criteria could be used and there is no record in the Council's literature of them actually being used. In the case of the intense neutron generator (ING) decision, the Science Council evaluated the proposal for this $100 million facility under the following headings:

- Scientific and technological significance
  - (a) significance to Atomic Energy of Canada Ltd
  - (b) significance to the scientific community
  - (c) significance to the universities
  - (d) significance to industry
  - (e) significance to the nation and public.
- Technical feasibility
- Manpower considerations
- Costs
- Financial scale (effects on Atomic Energy of Canada and other R & D performers)
- Organization
- Location
- Alternatives to the ING.

Elements of 'Weinberg's criteria' are contained in this evaluation though there is no specific citation to his work in the ING report.[12] No attempt was made to quantify the various criteria, since the proposal was not being judged competitively against others. The general opinion within the Science Council now is that its advice to the government to proceed with the ING was not sound. In reaching its conclusions, the Council relied too heavily for its advice on those people who had most to gain from the implementation of the proposal.[13]

THE DEPARTMENTS AND RESEARCH COUNCILS

Many federal departments carry out research and support it extra-murally through the contracting out procedure known as Make-or-Buy or by direct support through various R & D incentive programs since the early 1960s. The Make-or-Buy contracting out procedure was established in the early 1970s in response to the constant complaint made by the various committees inquiring into the Canadian research system that the Federal Government was too heavily involved in intramural R & D. In 1972, the Minister of State for Science and Technology announced that the government had adopted a general policy of looking to the private sector to satisfy its R & D needs. The Cabinet instructed the Treasury Board secretariat to provide guidelines to the departments for implementing this policy decision. The guidelines which eventually emerged designated the Department of Supply and

Services as the responsible agency for what came to be known as Make-or-Buy. The Department was also given the responsibility for determining the research capabilities of Canadian industry and for communicating the government's research requirements to potential contractors. The MOSST, in consultation with the departments and agencies, was asked to take responsibility for reviewing the contracting out program and reporting its findings from time to time. Implementation of Make-or-Buy began in 1972, applying initially to all new projects within the participating departments, excluding sensitive projects in the defence and regulatory fields. In 1974 the government expanded the original policy to cover unsolicited proposals from the private sector, a move that was intended to permit greater industrial participation in government science programs provided that there was a department or agency with an appropriate mission to sponsor the proposals.

There is no doubt that Make-or-Buy has been a significant success although the term itself is no longer used. $15 million was set aside for unsolicited proposals in 1982-83, and payments to industry by the departments and agencies amounted to about $390 million. Not all of this was accounted for by contracting out, however. More than $200 million was spent on the various direct assistance schemes operated by the former Department of Industry, Trade and Commerce, the Department of Energy, Mines and Resources and the research councils. The schemes of the Department of Industry, Trade and Commerce (now the Department of Regional and Industrial Expansion) are very significant. In 1976 the Department asked a consultant to examine its numerous assistance programs to industry. In 1977 the government announced that seven of the eighteen programs studied would be reorganized as a single program — the Enterprise Development Program (EDP). The EDP combines the basic features of these programs and is designed to facilitate co-ordination amongst various forms of assistance, making the programs more accessible to Canadian industry, particularly small— and medium—sized firms. The EDP is administered by the Enterprise Development Board and the regional Enterprise Development Boards, all of which report to Cabinet through the Minister of Regional and Industrial Expansion.

The overall objective of the EDP is to enhance the growth of the manufacturing and processing sectors of Canadian industry by providing assistance to selected firms to make them more viable and internationally competitive. It is also intended to foster innovation and smooth adjustment to increasingly competitive markets in Canada and abroad. The focus for assistance is on promising smaller and medium sized firms prepared to undertake projects which are high-risk in relation to their resources, which are viable and which promise attractive rates of return on the total investment. Companies that can demonstrate the need for financial assistance and the capability to pursue successful R & D projects are eligible under the EDP for cost-shared assistance for process and product innovation and market research. Funding of up to 75 per cent of eligible costs is available, these costs being defined as

- research, development and design projects provided these represent a significant burden on the company's resources;

- engagement of consultants for market feasibility, product enhancement, product development and design studies, and pollution control and abatement projects.

Expenditure on the EDP doubled between 1979-80 and in 1982-83 $94 million were allocated.

The Defence Industry Productivity (DIP) program assists, develops and sustains the technological capabilities of Canadian defence industries and encourages consequent defence export sales. In 1982-83 about $66 million was set aside for direct assistance to these industries for product development, pre-production expenses and capital costs on a shared basis.

Many federal departments carry out and support extramurally a range of R & D activities (see Table 7.4). There are certain complex and important areas that cross departmental boundaries, however, and special programs of research have been inaugurated in recent years. To date, thirteen of these special applications areas have been identified. The programs are controlled by Interdepartmental Committees (IDCs) which ensure co-ordination and liaison between the departments and agencies concerned. These IDCs have emerged independently and at different times and have mandates to meet the differing needs of each area. Secretariats are usually provided by 'lead' departments; for example, the energy program's lead department is Energy, Mines and Resources; the lead department for the space program is the Department of Communications. Lead departments within an IDC framework are required to report to Cabinet and the Treasury Board on the utilization of existing resources and to recommend changes of priority for new or existing resources where appropriate. Whereas the Transportation and Energy IDCs can influence budgetary decisions, the others have co-ordinating functions only and do not make recommendations on resource allocation to programs. It appears that these priority areas have not emerged as a result of any sophisticated method of priority assessment. They have appeared as departmental activities in the areas have increased, and as the areas have become important both nationally and internationally.

THE RESEARCH COUNCILS

There are three research councils in the federal research system, the

Medical Research Council (MRC), the National Research Council (NRC) and the Natural Sciences and Engineering Research Council (NSERC). The oldest is the NRC which was established in 1916 following the creation of DSIR in Britain. All three are Crown corporations and report directly to a designated minister. The MRC is the smallest with a budget of $113 million in 1982-83. Founded in 1938, it became a Crown corporation in 1969 and now reports to the Minister for National Health and Welfare. About 97 per cent of the Council's

Table 7.4

Federal expenditures on the natural and human sciences

Department	($ million)	
	1981/82	1982/83
Total science	2604.6	2941.5
Economic development	1511.2	1754.7
Agriculture	168.5	196.1
Communications	81.3	66.7
Energy, Mines and Resources	229.7	280.5
Atomic Energy of Canada Ltd.	114.2	132.4
Environment Canada		
Forestry Service	58.3	69.4
Fisheries and Oceans	130.0	145.8
Industry, Trade and Commerce	143.5	173.3
National Research Council	297.4	360.7
Natural Sciences and Engineering		
Research Council	201.8	227.1
Supply and Services	15.1	15.1
Transport	21.4	24.0
Others	50.0	63.6
Social Development	604.4	693.4
National Library	21.6	25.8
National Museums of Canada	58.7	61.8
Social Sciences and Humanities		
Research Council	46.6	56.7
Employment and Immigration	18.9	18.3
Environment	220.7	264.7
National Health and Welfare	72.6	80.9
Medical Research Council	100.2	112.9
Others	65.1	72.3
External Affairs and Defence	207.9	251.4
CIDA	40.4	45.0
IRDC	46.0	56.6
National Defence	112.0	139.6
Others	9.5	10.2
Services to government	281.1	242.0
Statistics Canada	230.0	187.2
Treasury Board	12.9	15.4
Others	38.2	39.4

Source: *The Government of Canada's Investment in Science* (Ottawa, 1982)

budget is used to support research, research training and related scientific activities in Canadian universities.

The National Research Council

The National Research Council (NRC) was created by Order-in-Council in 1917 in response to wartime needs for rationalizing Canadian scientific research. Initially, it was an honorary advisory body to the Privy Council committee on scientific and industrial research, a Cabinet committee charged with advising the government on all aspects of the planning and co-ordination of its scientific activities. At this time the principal aim of the NRC was to promote industrial R & D. In its early years considerable time and effort went into establishing this body of scientists as a powerful voice in the federal political structure. It built up a program of support for research in universities and assisted, through some twenty-five associate committees, a number of industrial research programs. Between 1930 and 1939 it began to establish its own research establishments and in the process it compromised its governmental advisory role. The Royal Commission on Government Organization of 1963 (The Glassco Commission) outlined the situation as it saw it at that time.

> The system has failed to function as intended. The Privy Council Committee has met infrequently and between 1950 and 1958 was not called together at all. The National Research Council has turned aside from its original duty of advising on broad national policy and has concentrated its efforts, albeit with conspicuous success, on the support of research and scholarship in the universities and, in a general way, on its own laboratories and establishments and the fields of science in which they operate.[14]

As a result, the Commission felt that decisions about science tended to fall by default on the Treasury Board. Furthermore, in the post-war years there was less disposition on the part of NRC to follow instructions that it had itself given to its own advisory committees (the associate committees), namely to look for that sector (industry, university or government) that could best perform the research. In a period of general expansion it became relatively easy to start new projects on the initiative of the working divisions of NRC and carry them out there. Priority reviews did not constitute any formal evaluation of past and present programs nor did they question, or accomplish any significant changes in, the decentralized system of NRC operation. Not only had the advisory function been displaced, the aim of developing industrial research also appeared to be displaced. The combination of the growth of the laboratories and their wartime experience with in-house initiatives tended to linger in the 1950s and this was reflected in a greater tendency to assume that each research priority and the facilities that it needed ought to be located at the centre in NRC laboratories.

In 1968 the NRC laboratories were restructured to overcome the problem of internal evaluation and how to provide for it organizationally. Positions were created for a Vice-President for Industry and a Delegue-general, a senior scientist with responsibility for advising on long-term objectives and methods of evaluation of objectives and performance.

When the Lamontagne Committee began its hearings and investigations the NRC became a prime target. The Committee came to the conclusion that there had been an undue reliance on in-house research by the Canadian Government for too long and that the NRC was one of the main instruments of this policy. The Committee recommended that the NRC should become an Academy, devoting all its activities to basic research and long-term applied research with no specific objective. Most of the government's intramural basic research would be concentrated in the NRC, some of it on contract from government departments and the universities. To complement the proposed Academy, the Committee advocated the establishment of a Canadian Industrial Laboratories Corporation (CILC) to serve the needs of the manufacturing sector. The Committee's recommendation for the establishment of an Academy out of a dismembered NRC was never adopted. Neither the Academy nor the CILC was established. In 1978, however, a new Natural Sciences and Engineering Research Council (NSERC) took over the grants and scholarships function of the NRC. This was not the first time that the NRC had lost control over major activities. The Defence Research Board, Atomic Energy of Canada Limited and the Medical Research Council had all at one time been NRC responsibilities, but loss of the granting function cut the Council off from the development of basic scientific research in the universities, an area of responsibility that it had had since its foundation and which was in keeping with the ethos of the organization. Henceforward it was clear that there would have to be a much greater emphasis on research applications rather than on the overall development of the national scientific research effort.

Priorities in the National Research Council

In 1980 the Council prepared its first long-range plan which spelled out Canadian economic and social problems and delineated the NRC's role in helping to solve them. In this impressive long-range plan the main problems were identified as the huge federal deficit and adverse trade balance, the high inflation rate and relatively high unemployment rate. Important factors identified as responsible for the situation included a relatively small manufacturing sector performing comparatively little technical research and a low gross national expenditure on R & D. In helping to relieve the adverse economic situation, NRC would organize itself around the following six themes:

1 Industrial expansion: the undertaking of science and technology initiatives with the objective of doubling the output of the techno-logy-intensive industrial sector by 1990;

2 Regional development: the development of science and technology programs in co-operation with provincial agencies in many areas of technology such as oceans, energy, food and manufacturing, in order to reduce regional differences;

3 Energy: expansion of energy R & D, especially in alternative sources and conservation;

4 Core support: increase of NRC competence in core technologies and related scientific disciplines so that this essential activity — the base for the first three themes — is restored to 30 per cent of NRC effort;

5 Social impact: the maintaining of current levels of scientific and engineering research in support of social progress in security, safety, health and environment in order to take advantage of new tech-nology and optimize the social effects of the first three themes;

6 Effectiveness: improvement of the NRC's effectiveness by consulta-tion, greater provincial liaison, increased industrial contact, increased public awareness, more complete ancillary services and improved exploitation of international exchanges.

The resources needed by the NRC to service the research programs resulting from these themes would rise from the 1980 level of $240 million to between $700-$1250 million.

Despite the emphasis on economic and social criteria, however, the NRC ethos noted in Chapter 3 would not be lost. In his introduction to the long-range plan, the President said:

> The wise husbanding of these resources and the vital urgency of the economic challenge will require all of the NRC's traditional flexibility and short response time, plus greatly increased inter-agency collaboration and a renewed sense of urgency. At the same time, the Council will be providing for the nation one of the more important facets of our civili-zation: a milieu for intense intellectual experiences and the advancement of knowledge.[15]

The six broad themes that were set out as guidelines for NRC opera-tions in the 1980-1990 period were translated in the long-range plan into objectives, research strategies and R & D programs. The objectives derived primarily by 'an analytical evaluation process' are shown in Table 7.5. Research programs and strategies were synthesized from consultations with laboratory and program directors and with indivi-dual scientists, engineers and others within the NRC itself. The various programs were costed using three alternative plans, all of which gave

Table 7.5
Themes and their objectives guiding NRC operations

Theme	Objective
1	To devise and implement science and technology initiatives that will contribute to a doubling of the output of the technology intensive sector by 1990.
2	To improve significantly the economic performance of selected regions by co-operative science and technology programs with the provinces and with other regional research organizations.
3	To participate fully in Canada's goal of achieving energy flexibility by developing appropriate science and technology programs.
4	To increase the component of internal NRC R & D that is devoted to core technologies and essential scientific disciplines from the current level of 25% to 30% by 1985-86. This activity is the foundation for themes 1, 2 and 3.
5	To provide appropriate support to optimize the effects of themes 1, 2 and 3 on such areas as health, safety, environmental quality and security and to conduct R & D in these areas.
6	To multiply the results of the NRC's programs by promoting and marketing them to a much greater extent and by providing ancillary services.

Source: The Urgent Investment: A Long Range Plan for the National Research Council of Canada (Ottawa, 1980)

first priority to the technology intensive industrial sector. The first alternative (flexible resource level) assumes that the NRC long range plan will be implemented as proposed ($1250 million at 1980 prices). The second alternative (identified resource level) assumes that most of the initiatives will be implemented, but no allowance is made for new proposals until five years into the Plan, and also a smaller involvement by NRC personnel in administering the extramural programs demanded by the technology intensive industrial sector. Under this alternative the financial resources needed would be $870 million. The third alternative takes the gloomy view that the programs cannot be sustained at the levels proposed in the second alternative and new projects either severely curtailed or dropped. This alternative would cost $700 million at 1980 prices.

In its plan the NRC is concerned to show that it is responsive to the social and economic needs of the Canadian people and has oriented its activities towards these needs. As evidence, it points to the direct assistance programs to industrial R & D that it pioneered, to the provision of consulting, information and other services to industry and to the increased contracting out that has been NRC policy in recent years. In 1974 only $2 million was contracted out to industry by the NRC. The annual figure now stands at more than $80 million. The NRC's direct assistance schemes for industrial R & D are of particular note. The Industrial Research Assistance Program (IRAP) was established in 1962 as a civil counterpart to the Defence Industry Productivity Program (DIPP) administered by the Department of Industry, Trade and Commerce. The general objective of the IRAP is to stimulate the establishment of new industrial research groups, or the expansion of existing ones. The government shares the cost of a project by paying the salaries and wages of R & D workers. It is generally believed that the program has increased the technological capability of Canadian companies and has generated significant economic benefits from the relatively small investment which, in 1982, amounted to about $23 million. The Council's Program for Industry/ Laboratory Projects (PILP) was introduced in 1975. Its purpose is to help Canadian companies to develop projects that the NRC itself has initiated to the point of exploitation. A small proportion of projects has now reached the marketing stage. The scheme was extended in 1978 to include worthwhile projects from other government departments. This was designated the Co-operative Projects with Industry (COPI) program. Some two years later the NRC took over the management of the consolidated PILP and COPI schemes which, in the 1981-82 budget were allocated almost $13 million.

The Natural Sciences and Engineering Research Council

The Natural Sciences and Engineering Research Council (NSERC) was established in 1978 to 'promote and assist research in the natural sciences and engineering and to advise the designated minister on matters relating to such research as may be referred to it by him for consideration'. The Council is a Crown corporation, governed by the President and a Board of twenty-one members. Its main activity is the awarding of grants and scholarships to university research workers. At the present time, NSERC perceives itself as an important instrument of government policy on several counts. It is helping to ensure that an adequate supply of highly qualified manpower is available to support the government's stated policy that GNERD should be 1.5% of GNP by 1985. Without the NSERC's support schemes this could only be achieved by importing the necessary expertise. It assists the government in its economic policies by providing the means to increase knowledge and develop problem-solving ability. The NSERC is in no doubt that the link between R & D intensity in an industry is directly related to productivity, real output and price competitiveness. The NSERC, like the NRC, promotes civilization through science because

> science is a characteristic and essential component of our civilization. It is an aesthetic intellectual and cultural adventure worth doing for its own sake. The study of 'black holes' in space or the general theory of relativity may never produce major economic benefits but to see through them some glimpse of the nature of our universe is worth as much as material benefits.[16]

The NSERC distributes its funds under four headings. Peer-adjudicated grants are awarded to university research workers for what is described as 'free' research, on the basis of peer review. Under the highly qualified manpower training and development scheme, scholarships and fellowships are awarded in national competitions to postgraduate students, postdoctoral fellows and senior scientists in universities, industrial firms and other institutions in Canada and abroad. The national and international activities program provides grants to support national and international conferences and studies, promotes the international exchange of scientists and engineers and supports the activities of scientific and learned societies. All of these activities are oriented more towards 'free' research than they are towards what the Council describes as 'targeted' research. In the latter area, developmental grants are the instrument of policy. These are designed to encourage university research workers to undertake research in selected priority areas of national concern, and are divided into two schemes. In 1972, on the assumption that Canadian university research is capable of making significant contributions to the development of Canadian industry, the NRC established the Project Research Applicable to Industry (PRAI) grant scheme, the objectives of which were

- to identify new research advances in university laboratories that have strong industrial significance and a high probability of benefit to Canadian industry;

- to support the development of these advances undertaken in collaboration with Canadian companies to the stage at which the results can be exploited by Canadian industry;

- to increase the benefits derived from the NSERC's regular programs of grants to individuals and negotiated grants without undue distortion of the individual's longer-term research interest and capacity for further development.

Any member of the academic staff of a Canadian university who is eligible to apply for an NSERC peer-adjudicated grant is also eligible to apply for a PRAI grant. A prerequisite for support under the PRAI program is that the project be undertaken in collaboration with a company to ensure that it meets a definite industrial need or opportunity in Canada. In return for its support and participation the collaborating company normally gains the advantage of first right of access to the results of the exploitation. Selection criteria are as follows:

- nature of the project: PRAI grants provide short-term complementary support for those whose research has already led to the identification of a specific and novel technique, process or product which promises to be of commercial value to Canadian industry;

- probability of technical success;

- probability of commercial success;

- economic benefit to Canada.

The grants are normally for two years or less. Estimated expenditure on PRAI grants in 1981-82 was $2.2 million, and by 1984-85 it is expected that $3.5 million will be allocated.

In mid-1977 the NRC announced a new strategic research grants program of support for university researchers working in specific areas of national concern. This was a direct result of a government request which said

> In terms of national problem areas, the government has in mind that the councils in their granting policies should give special emphasis to the improvement of our knowledge of the nature and scope of these problem areas and of our capability to find solutions to them. Examples of areas of such national concern are energy conservation and supply, transportation and communications...

The NRC decided on three priority areas—energy, environmental toxicology and oceanography, and invited submissions from university scientists. The response to this new scheme was good. It was significantly better than the response to the PRAI scheme, despite misgivings within the scientific community that the program would operate at the expense of the peer-adjudicated grants scheme. In 1981-82 the budget for the strategic research grants program was $20.6 million, almost double the figure for 1979-80. The NSERC has added Communication and Food/Agriculture to the list of priority areas.

CO-ORDINATION

The main thrust of Canadian science policy in the last two decades has been towards an improvement of the nation's technological innovation, a lowering of the government's involvement in R & D and a reduction of Canada's technological dependence on other countries, especially the United States. The great number of industrial R & D incentives schemes that have been generated is indicative of the seriousness with which these priorities have been pursued. There has been a continuing concern about the country's low position on the international league table of expenditure on R & D and a major national objective is to increase Canada's GNERD to 1.5 per cent of the GNP by 1985.

The science policy machinery that has been instituted to achieve these objectives is based on the concerted action model. There is a science policy minister who, in theory at least, has a co-ordinating role and there is an external advisory body. Research and development is carried out in and funded by a number of departments and crown corporations whose budgets are, in principle, subject to vetting by the Minister of State for Science and Technology. The existence of an *ex ante* science budget is central to the concerted action concept but in Canada this has not yet been achieved. In 1979 the MOSST proposed to the Cabinet that it should have a more active (rather than reactive) role in the budget process by placing before the appropriate Cabinet Committee details of the size of the proposed science budget, the preferred areas of science and technology and proposed sectors to perform the work. It was intended that if Cabinet agreed with these proposals interdepartmental co-ordinating panels similar to the existing ones in space, energy, oceans and transportation would be set up. At this stage the MOSST consults with the departments and statutory corporations at an earlier point in the budget process than before and it has acquired their co-operation in the compilation of *ex post facto* science and technology expenditure statistics, which it produces very quickly after the budget.

Japan

Japan has been transformed in less than a century from a weak, isolated, agrarian state to an industrial giant, ranking third in the world's list of industrial producers. After the Second World War, when the country's economy and most of its industrial plant lay in ruins, two strategies were put forward for economic recovery. The first, which was advocated by orthodox financiers and economists, favoured the development of labour-intensive industries which would soak up the great pool of unemployed. The second, advanced by bureaucrats and administrators in government, favoured the importation and adaptation of overseas technology and the pursuit of technical excellence and innovation in the processes of production. For more than thirty years after the war's end it was the second strategy that was pursued, and with great success. Industries with world market potential and the potential for reduction in production costs through economies of scale were selected for development. These industries were carefully protected in their infancy by tariff and non-tariff barriers and eventually left to fend for themselves against outside competition.

The success of Japan's industrial strategy was not due solely to the skill showed by the decision-makers in choosing industries for development, however. It was due, also, to the superiority of management, harmony between workers and management and a degree of co-operation and coincidence of aims between government and industry that is unusual by Western standards. The role of the state was crucial to Japan's post-war economic recovery and this could only have taken place against a background of such co-operation.

Paralleling the growth in the Japanese economy in the 1950s and 1960s was a substantial growth in its R & D expenditure which, by 1968, stood at 1.5 per cent of the GNP. By 1978 it was 2.18 per cent of GNP and 98 per cent of this was being spent in the civil sector. This

contrasts with British and American civil R & D expenditure which
accounts for about 48 per cent of the overall amount spent. Figure 7.3
shows GNERD figures for various countries, from which it will be seen
that Japan actually spent more on R & D in 1978 than the United
Kingdom, France and West Germany. And, as Figure 7.4 shows, the
private sector accounted for more than 60 per cent of all expenditure
on R & D. In the light of such dominance, it is surprising to find that
the influence of the government on the direction taken by the
industrial sector has been, and still is, so great; of all the advanced
industrial nations Japan has the smallest public sector and the lowest
ratio of public expenditure to GNP and yet the government and the
bureaucracy can wield enormous influence. Allen has explained the
influence of government thus:

> In Britain ... the interests and purposes of government and private enter-
> prise were different. Politicians and civil servants occupied different camps
> from the industrialists. The latter might call on the government at times to
> defend their interests, but in general they regarded government as a power

Figure 7.3

R & D expenditure of selected countries

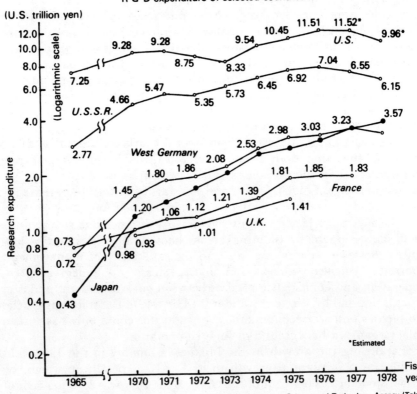

Source: A Summary of 1979 White Paper on Science and Technology by the Science and Technology Agency (Tokyo,
1980)

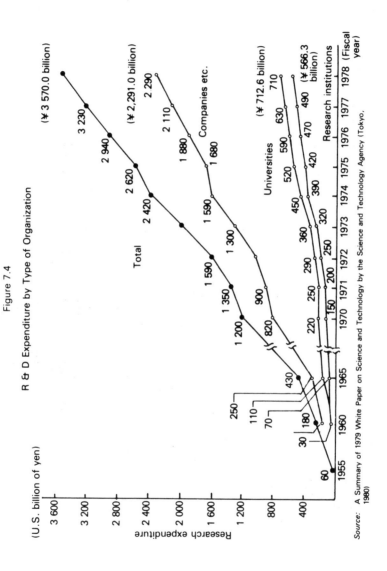

Figure 7.4

R & D Expenditure by Type of Organization

Source: A Summary of 1979 White Paper on Science and Technology by the Science and Technology Agency (Tokyo, 1980)

that curbed and frustrated their activities. In Japan this dichotomy was
absent. Indeed it was alien to the Confucian tradition inherited from times
long past. In general, policy making has been regarded as a combined
operation in which both government and industry were aiming at the same
target—rapid industrial growth.[17]

Where necessary, industry can be influenced more directly than by the
discussions leading to consensus. The powerful Ministry for International Trade and Industry (MITI), for example, periodically issues
administrative guidelines—directives which lack legal status but are
rarely ignored by the private sector.

So the Japanese economic recovery has not depended on pioneering
R & D leading to new products. But this situation is now changing.
More Japanese patents are now registered abroad than foreign patents
are registered in Japan and though the adverse technological balance
of payments is still about US$900 million, this is mainly due to pay-
ments on previously imported technology. In other words, the strategy
of importing technology from overseas is on the wane and Japan has
joined the United States and Europe at the frontiers of R & D.

Dixon Long distinguishes two tiers in the Japanese system for the
management of its governmental research. On the 'staff' tier there are
the central policy co-ordinating and advisory bodies—the Council for
Science and Technology, the Science and Technology Agency and the
Japan Science Council. On the 'line' tier there are the executive ministries with their own advisory and co-ordinating machineries (see Figure
7.5). Various ministries and agencies are involved in funding and conducting R & D but two, the Science and Technology Agency, which is
both an executive and an advisory body, and the Ministry of International Trade and Industry, account for over 70 per cent of the
government's R & D commitment between them.

THE COUNCIL FOR SCIENCE AND TECHNOLOGY

The Council for Science and Technology is a central Cabinet-level
policy advisory body located within the Office of the Prime Minister.
Its members include the Prime Minister as chairman and the Ministers
for Education, Economic Planning, Science and Technology, and
Finance. In addition, the membership includes the President of the
Japan Science Council and five Prime Ministerial appointees who have
to be approved by both Houses of the Parliament. The Council co-
ordinates the administrative policies of all government agencies by
establishing fundamental and comprehensive policies on science and
technology, establishing long-range and comprehensive goals for the
nation's science and technology effort and formulating policies for the
research that is important in attaining these goals. In formulating
policy the Council for Science and Technology receives advice from the
Japan Science Council and other advisory bodies within the Office of

Figure 7.5

Main elements of the Japanese government's advisory, executive and funding arrangements for R & D

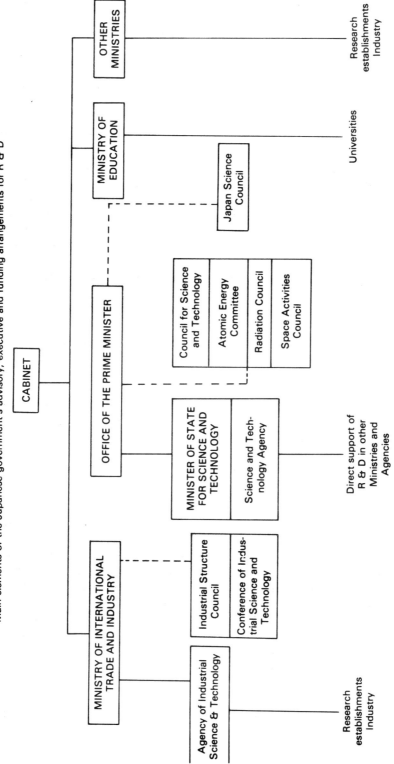

the Prime Minister. The statutory membership of the President of the Japan Science Council and the existence of a large and complex sub-committee for co-ordination between the two Councils emphasizes the importance that is placed on the acquisition of independent advice. Other specialized advisory councils — for space, atomic energy, ocean development and radiation — are brought into a working relationship with the Council through cross-membership of one or more of their appointed members and of Cabinet officers. The Science and Technology Agency is a particularly important source of advice on policy and co-ordination, and it provides a secretariat for the Council.

THE JAPAN SCIENCE COUNCIL

The Japan Science Council was established in 1948. It has 210 members, 30 in each of its seven divisions, and a network of committees ensures contact with all parts of the country. The specific duties of the Council are:

- to deliberate on and assist in the realization of important matters concerning science;
- to promote and maintain liaison between research workers;
- to encourage efficiency in research activities.

It is also required to advise the government in the following areas:

- promotion of science and the development of technology;
- improvement of the utilization of results of research;
- training of research workers;
- the place of science in state administration;
- promotion of science into national life in general and into industry in particular.

The Council is empowered to act independently in co-ordinating research for greater efficiency. Importantly, it has to advise on the state R & D budget and on budgets for institutes and laboratories under the government's control.

Dixon Long has characterized the historical relationship between the Japan Science Council and the Council for Science and Technology as a contest which has ended in an uneasy compromise in which the Council for Science and Technology has kept its hands off research in the education sector and the Japan Science Council, despite its impressive terms of reference, has had to content itself with making recommendations relating to pure research and with having some modest operational responsibilities in the international science policy sphere.[18]

THE SCIENCE AND TECHNOLOGY AGENCY

The Science and Technology Agency (STA) is located within the Office of the Prime Minister and is headed by a Minister (Director-General) of Cabinet rank. It originated in the Scientific and Technical Administration Committee, created in 1948 as a co-ordinating body for government R & D activities. In 1956 the Atomic Energy Commission and several other advisory and administrative bodies were added and the STA came into being as an administrative organization to co-ordinate the government's research efforts. Its duties are to

- plan and implement basic policies;
- co-ordinate the R & D of ministries and state agencies, with the exception of MITI, in order to increase effectiveness and avoid overlaps;
- promote national projects in new fields requiring quick and energetic exploration;
- implement basic research in areas common to several disciplines;
- promote work on atomic energy.

The STA is thus an important executive and advisory body tendering its advice directly to the Prime Minister's office. As an administrative subsidiary of the Office of the Prime Minister it connects the advisory and executive elements of the governmental science policy machinery. Through the functionally-organized bureaux, it links the administration of major government R & D programs with their policy advisers and, through its Planning Bureau, provides a link with the Council for Science and Technology. It has an important co-ordinating function in the R & D budget, as is shown by the following budget sequence:

(i)   research and development activities are proposed by the laboratories and public corporations under the aegis of particular ministries and agencies;

(ii)   particular items for co-ordinated research or block funding are negotiated jointly by STA and the ministry/agency concerned;

(iii) ministries and agencies submit budget requests to the Ministry of Finance, including R & D items;

(iv)   research and development items (except those of MITI and the Education Ministry) are packaged for review by STA;

(v)   authorization from the Ministry of Finance following STA review.

The STA is also involved in promoting industrial research and is responsible for the Research Development Corporation of Japan, set up in 1961 to select and contract to industry development projects of promise

but too large, risky or long-term for private industry to undertake un-
aided. Promising projects are submitted to the Corporation by
national laboratories and those selected are put out to tender. A
successful project is developed and exploited by the chosen firm, which
repays the Corporation over five years, and the firm enjoys monopoly
rights for an agreed limited period. The STA also promotes inventions
by subsidy and is responsible for the Japan Information Centre of
Science and Technology, the services of which are heavily used by
industry, and it operates a number of large national laboratories. It
relies on its own network of advisers and consultants and on government guidelines in setting its priorities. Each year the Planning Bureau
issues a White Paper which serves as the guideline for priority-setting
in the year ahead. In the 1980 White Paper, for example, the development of Japan's technology, especially in energy production, electronics, health, materials, transport, space and the oceans, was regarded as
having priority in the 1980s.

THE MINISTRY OF INTERNATIONAL TRADE AND INDUSTRY

The Ministry of International Trade and Industry (MITI) is the
ministry with primary responsibility for administering government
policy towards industry. All state laboratories concerned with industrial R & D are controlled by the MITI and are administered by its
Agency of Industrial Science and Technology (AIST). The latter is not
only responsible for the direct administration of R & D in the laboratories under its control but it is also responsible for encouraging
research in private laboratories. It is aided in this by various advisory
councils. The Industry Structure Council integrates the recommendations of all specialized councils into an industrial strategy. It advises the
Industry Policy Bureau of the MITI, and its recommendations form
the basis of future orientations of R & D. Its mandate is to define the
general orientation that national industry is to take in the light of
social, economic and political circumstances, both domestic and international. The thirty members include representatives from industry,
commerce, universities and the labour unions. The Industrial Technology Council, comprising experts from industry, the universities and
government laboratories, advises the AIST on areas of R & D that
should be explored in line with the recommendations of the Industry
Structure Council. Its eight committees define and monitor a range of
research programs that the AIST carries out in association with the
private sector. The Conference of Industrial Science and Technology
has a membership of 200 scientific organizations which break down
into committees for the preliminary assessment of technological developments and the definition of comprehensive and long-term R & D

objectives in specific areas defined by the other advisory councils.
The specific responsibilities of AIST are:

- the development and utilization of new energy sources, to diversify
current energy sources and to develop new energy conservation
technology;
- oversight of the National Research and Development Program
which was initiated in 1966 and is financed by the Treasury with the
co-operation of industry and academic institutions. Projects funded
under the program must be of urgent national importance in improving national industrial structure, promoting efficient use of
natural resources and preventing pollution; expected to accelerate
technological progress, especially in mining and manufacturing;
outside the scope of private industry as too costly, long-term or risky;
clearly specified and examined, and carried out in co-operation with
government, universities and industry;
- promotion of technology for health and welfare; construction of
Tsukaba research centre to create a favourable research environment; promotion of international research co-operation; co-ordination of R & D conducted in AIST laboratories; establishment of
long-range strategy for R & D and industrial technology; research on
industrial technology information systems; promotion of technology
assessment; development of computer technology; promotion of
private industrial R & D;
- operation of 16 national and 195 prefectural laboratories.

Priority-setting in MITI has been described as 'a set of consensus-
oriented pyramidal networks operating in succession . . . The network
with the broadest perspective takes precedence and lays down constraints within which the networks with more specific interests must
expand upon policies and implement them.'[19] The first stage is the
drafting of non-technological national needs, which are defined as
basic constraints on national technology policy. The next stage involves
a broad technical approach in which industry, universities and government departments are involved in the Conference on Science and
Technology. Finally, a specific and technical approach is adopted
which defines R & D programs to achieve the objectives and priorities
that have emerged at the higher levels, and advice is given within the
context of the Industrial Technology Council on the sharing of
research tasks between the various performance sectors. Technological
priorities are, therefore, determined through a system based on consultation between business, government, universities and consumer
representative bodies.

CO-ORDINATION

The Japanese R & D system is dominated by the private industrial sector, which funds more than 75 per cent of the country's research effort. National objectives have, in the past, been oriented towards achieving equality with the West and a vigorous program of technology importation has been in progress for many years. The introduction of foreign technology rapidly closed the technological gap between Japan and the West but the flow is now drying up and restrictions placed by foreign owners of technology are causing some problems. A major national objective now in force, therefore, is an increase in indigenous technological innovation to overcome the decline in availability of foreign technology and to help the country to cope with the difficulties imposed by the 'new economic context' of the 1980s.

The governmental science policy machinery leans towards the concerted action model. There is a central committee of ministers, the Council for Science and Technology, which is chaired by the Prime Minister, and this receives advice from sources both internal and external to the government. The Japanese Science Council is the main source of external advice and this is analogous to the independent advisory council of the concerted action system. The powerful Science and Technology Agency has both a co-ordinating and executive role, a departure from the concerted action system in the sense that it is not usual for the Ministry of Science Policy, which the Agency is to all intents and purposes, to have an operational role; as the OECD put it, the STA is both judge and advocate.[20] Ministries with the exception of the Ministry of Education submit their budgets to the STA for negotiation before they are sent to the Ministry of Finance. Priorities are determined by a widespread consultative process and there is no evidence that any formal qualitative techniques are used in setting priorities at any level.

CHAPTER 8

Science policy in Belgium

Towards a centralized system

The present science policy machinery in Belgium originated in the work of a National Commission set up in 1957 to study the problems posed by scientific progress and its social and economic repercussions. In its report, submitted to the government in 1959, the Commission recommended the establishment of an institutional science policy structure comprising a ministerial committee, an interministerial commission and an advisory council. This structure was instituted in that year and the general aims of Belgian science policy were set out at that time as:

- the maintenance of an adequate level of resources to R & D, compatible with budgetary constraints and national revenue;
- rationalization of the utilization of resources devoted to scientific activities. The focus here is on the co-ordination of institutions, and the harmonization of statutes as well as the reduction of duplication of research programs;
- the direction of general basic and applied research efforts of the nation in accordance with scientific needs but also in accordance with cultural, social and economic needs;
- the development of an adequate system for the processing and dissemination of scientific and technical knowledge;
- the promotion of the utilization of the results of research in the national interest;
- the preparation and co-ordination of Belgian positions regarding scientific and technological co-operation on the international level in order to properly integrate the Belgian industrial and scientific potential in the international programs.

Within the government there is now a Ministry of Science Policy with a powerful secretariat—the Science Policy and Programming Services (SPPS), a Ministerial Committee of Science Policy (CMPS) and an

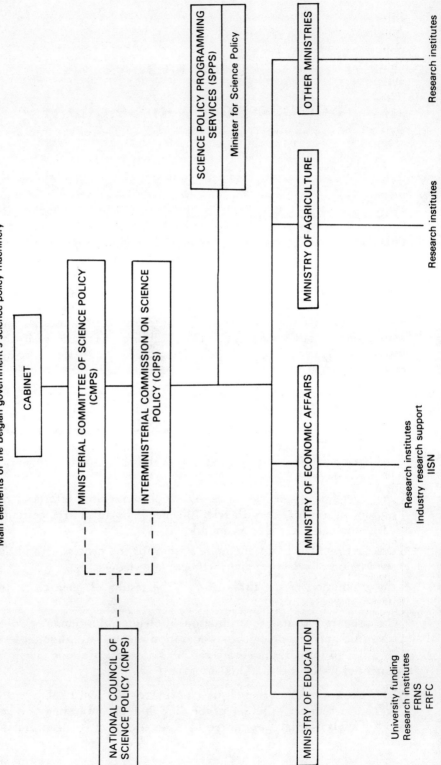

Figure 8.1

Main elements of the Belgian government's science policy machinery

Interministerial Commission for Science Policy (CIPS), representing the main executive and co-ordinating elements of the nation's science policy machinery (see Figure 8.1). The Minister of Science Policy is charged with the overall co-ordination of the government's scientific activities and he does this at the decision-making level within the CIPS. The National Council of Science Policy (CNPS) acts as an independent advisory body to the government, forming the final link in the chain that makes up this particular model of science policy machinery.

The Ministerial Committee of Science Policy (CMPS), presided over by the Prime Minister, is responsible for formulating and executing Belgium's science policy and for co-ordinating the work of the departments involved in R & D and related activities. Its membership includes the Deputy Prime Minister, the Minister of Science Policy, the Secretaries of State for French and Dutch Culture, the Minister for Finance and ministers whose departments have significant responsibilities in science and technology. The Committee has the final responsibility for the Science Budget before it is put to the Cabinet. The Interministerial Commission for Science Policy prepares the files of the ministers involved in the activities of the CMPS. It is a top-level co-ordinating body of senior public servants from these ministries and is presided over by the Secretary-General of the Science Policy Programming Services, with which it shares a secretariat. The CIPS, on instructions from the CMPS, co-ordinates and executes government decisions on science and technology policy when they involve joint action by more than one government department.

The Science Policy Programming Services

The Science Policy Programming Services (SPPS) is a specialized administrative body within the Department of the Prime Minister, coming under the jurisdiction of the Minister of Science Policy. The Secretary-General is its permanent head and its main tasks are:

• to co-ordinate all university research activities in the country;

• to co-ordinate the research activities of the government departments;

• to provide secretariat facilities for the National Council of Science Policy as well as chairmanship, secretariat and administrative support to the CMPS.

Apart from these administrative and co-ordinating functions, the SPPS has other information collection and support functions that are listed under four general headings.

General activities
Inventory of the national science and technology potential, involving
a biannual survey of the whole of the Belgian research effort, both
public and private.

Preparation of the science policy budget.

Programming activities
Studies of university growth, involving manpower planning within
subject areas so that the growth of the university and higher education
sector can be monitored and remedial action taken at the budgetary
level if this is necessary.

Oversight of the 'parallel funding' of university research, involving
the processing and classification of available information on the research projects separately financed by the government in the universities so that this 'parallel' funding can be continuously monitored and
rationalized.

Oversight of the national system of technological development and
research, involving the collection of available information on public
sector R & D so that duplication can be eliminated, objectives set, and
the optimum balance between basic and applied research maintained
and criteria established for identifying priority areas.

Management of the space sector
The SPPS contributes its technical and administrative skills to the task
of managing the national space research program which is operated
largely within the European international space agencies. This is
effected with the assistance of the Defence Research Centre of the
Ministry of National Defence.

Co-ordination of national research programs
The SPPS has the responsibility for developing a better scientific or
technological understanding of problems that are of interest to a
number of departments, and which go beyond the limits of the departmental research budgets. With the co-operation of the departments
concerned, the SPPS defines the research needs and identifies the units
in the university, public or private sectors which are capable of developing research in the fields in question. Having identified these units,
the SPPS establishes national research programs financed by 'credits d'
impulsion' of the Prime Minister's budget established for this purpose.
This activity is carried out in the context of the Interministerial Commission of Scientific Policy. Various national R & D programs have
been started up or are being studied with the aim of

- attaining well-defined national objectives in the areas of health, environment and social progress;
- promoting an interdisciplinary approach to physical, biological and social phenomena;
- making use of scientific and technological forecasting;
- contributing governmental support continuously from research through to economic and social implications of the findings.

The science budget

It was pointed out in Chapter 3 that in a concerted action or centralized science policy system the science budget is a powerful instrument for implementing the policies that are formulated centrally. In Belgium, scientific activities are expected to fit in with the needs expressed in national economic and social plans but, unlike the more heavily centralized French science policy system, R & D plans are not regarded as a part of a national plan. This is because in Belgium the time-scale of R & D is, in general, regarded as more long-term than the economic and social concerns that appear in national plans. Nevertheless, the R & D plans of the various ministries are controlled by the budgetary process, although the general function of co-ordination is not meant to prejudice the powers and responsibilities proper to each of the ministries concerned in the management of the money supply and the organization of the institutions under their control. According to official documents, the science budget exposes duplications, allows a more accurate orientation of research activities to the functions they are supposed to serve and allows for the determination of priorities. As such it constitutes an essential instrument of policy in the original meaning of the word.

The science budget is formulated every year on the basis of the proposals from the ministries, which separate out their requests for funds for extramural and intramural support of R & D. The SPPS gathers detailed information about the amount, justification and assignment of these various requests and re-groups them by function to compare the overall amounts in particular sectors with government decisions on growth in these sectors and to list its views on the line items in the ministries' submissions. When the analyses are completed the budget proceeds to Cabinet through the CIPS and the CMPS.

Research in the universities

The Ministry of National Education provides direct subventions to the state and free universities, 25% of which are devoted to research. In addition, there are a number of 'foundations' for the support of university research—the National Foundation for Scientific Research (FNRS), the Fund for Collective Scientific Research (FRFC), the Fund for Medical Scientific Research (FRSM) and the Institute of Nuclear Sciences (IISN). These funds are generally restricted to fundamental research. The FNRS is the parent body and provides administrative support to the others. It is also responsible for the allocation of coveted research fellowships in the universities—more or less permanent posts.

The FRFC is a grant awarding body which, until recently, was the instrument for the implementation of important 'concerted action' research programs. In order to better define the aims and priorities of university research a decision was taken in 1970 to institute the first program of basic concerted research (Actions de recherche fondamentale concertées) and in 1971 150 million Belgian Francs were allocated to these research programs. The projects are chosen by consultation with the government on the one hand, which decides on appropriate areas of research, and the universities on the other hand, who propose specific projects and designate the research centre capable of carrying them out. The selection of the laboratories is made on the following criteria:

- the proposed project must be related to the sectors or scientific disciplines already represented in the institution;
- projects will ultimately lead to an exceptional advance in scientific knowledge in a particular field or will contribute to the solution of certain essential social problems;
- the centres must demonstrate that they have already achieved high quality research.

In the course of the first generation of concerted action programs (1970-75) about twenty laboratories or research centres were supported. From the scientific point of view the policy has achieved remarkable results which have, in certain respects, gone beyond the initial objectives. Existing or potential centres of excellence have been identified and a number of them developed to the point where they have achieved international reputations, and the program has provided for the perfection of technologies or new products capable of being exploited industrially or commercially.[1]

In 1976, however, changes were made which reflected some government dissatisfaction with the way in which the programs were responding to expectations, especially in the social science sphere. The administration of the programs was removed from the FRFC and the universities and placed in the hands of the SPPS. The institutions entitled to present concerted action programs, as well as the conditions these institutions have to fulfil in order to receive support, are now specified more closely. They are now required to establish a research committee which advises the directors of the institutions on the choice, justification and administration of the concerted action programs. The new arrangements provide for a progressive decrease in subsidy so that the institutions themselves will, as time goes on, support the programs to a greater extent from their own resources, and eventually fully support them. Thus, from 1976 onwards, the concerted action programs have been geared to the following objectives:

(i) the development within the university system of centres of excellence in fundamental research considered by each participating institution to have priority;

(ii) the development of inter-university centres of excellence;

(iii) the development of centres of excellence within universities integrating the carrying out of fundamental and applied research aimed at the social and economic utilization of the results of research.

The second generation of concerted action programs, which started in 1976, provides for the maintenance of twelve centres of excellence fulfilling objective (i), some of which have their origins in the first generation; under objective (ii) there are inter-university centres of excellence in molecular biology, catalysis and oceanography; under objective (iii) three centres of integrated research have been maintained from the first generation of the concerted action program.

The concerted action programs are an attempt by the Belgian Government to provide incentives for the development of centres of excellence and their use in the realization of socio-economic objectives. The funds for the concerted action programs are provided from the Prime Minister's Department and the SPPS plays an important role in the decision-making process and in monitoring the progress of the programs. Applications for support under this scheme are made to the Ministry of Science Policy.

The national research programs

Various ministries are responsible for the conduct of R & D in Belgium, and control over their activities is exercised through the science policy machinery. It has become apparent to the Belgian Government in recent years that many of the complex problems that it faces need an integrated approach over many fields, so it has recently established national programs of research to help it in deciding on technically feasible solutions. Given the present state of the development of science policy such instruments can be perfected provided that they 'direct towards clearly defined aims a sustained and programmed effort of R & D effected by organizations capable not only of an integrated scientific effort but capable also of constantly relating theory and practice.'[2]

The national research programs are drawn up by the SPPS in collaboration with the departments involved, discussed in the Ministerial Committee of Science Policy (CMPS), and approved by the Council of Ministers. These national programs constitute the 'public orders of research' in previously defined areas, their results are the property of the state and they are geared specifically to the needs of the ministries. Examples include a mathematical model of the North Sea and the Scheldt estuary, a map of heat sources and demands, and a model of the national energy system. The government expects that these programs will provide reliable scientific data to facilitate decisions relating to necessary investments either in capital or research. The programs also include research on new products, techniques, procedures or systems in fields which seem to have implications for particular government sectors, bearing in mind national strategic preoccupations and possibilities.

The elaboration of all national programs begins by an inventory of appropriate current research and an identification of the scientific bodies engaged in it. Once the programs are approved, the most qualified teams of research workers in universities and scientific establishments are invited to participate, and they are carried out under constant control by representatives of the departments who are directly interested. On the scientific level, a general co-ordinator and several subcoordinators monitor the programs and an operational director follows the day-to-day progress.

The National Research Program is a significant feature of Belgian science policy. It attempts to gather the diverse talents that exist in the national scientific research system around themes that are innovative, precisely defined and related to national needs. Government involvement in the program is intense in that the aims, the time constraints and the means of achieving the aims are defined beforehand and it is up to the government to effectively organize the use and transfer of the results that emerge. The programs now in operation are oriented to the problems of energy production, the rational utilization of primary resources and energy, the improvement of the environment, the improvement of the quality of life and the problems of health and social progress.

Industrial research support

Belgium is an importer of high technology and its policy over the past few years has been to decrease the level of its technological dependence in order to maintain a reasonably favourable balance between exports and imports. Various measures have been adopted in order to provide for its new industrial policy by creating a scientific and technological infrastructure appropriate to it. The Ministry of Economic Affairs is particularly active in providing R & D incentives to secondary industries, using mainly three policy instruments.

Government support to research associations and to projects in industry is given through the Institute for the Encouragement of Scientific Research in Industry and Agriculture (IRSIA), a body incorporated by law and located in the Ministry of Economic Affairs. Its aim is to promote, by direct subsidy, collective scientific and technological research. It acts as a major clearing house for information and is an important instrument for forging informal links between industry and the universities. The IRSIA has no laboratories of its own and generally funds about 10 per cent of all R & D in Belgium. Normally the IRSIA provides subsidies of up to 50 per cent for joint research programs put forward by industry. Projects may nevertheless be subsidized where one firm has a monopoly, and in the case of agricultural grants, the full cost of R & D projects is generally met. Grants are also awarded through the IRSIA to graduates studying for higher degrees in a university, or in an industrial/agricultural institution, and to research workers wishing to study in other countries.

Research projects may either be proposed to the IRSIA by an industrial or university research unit or they may be suggested to industry by the IRSIA. When the Institute receives a draft proposal, one of its technical advisers discusses it with the firm's scientific staff to determine the value of the project to other industries, the objective being to arrange a co-operative research program between two or more firms or a firm and a university. An example of such an initiative is the Comité Etude de l'Etat Solide, set up in 1957 with representatives from a number of large industrial laboratories engaged in solid state research and a university department. The IRSIA is generally considered to be an outstanding success. It brings industry and the universities together and pushes conservative industries into starting new projects, encouraging them to look further into the future and perform fundamental research beyond the level required for their individual needs.

The prototype service of the Ministry of Economic Development provides financial help for the development of prototypes, new processes and products by interest-free loans which are repayable after successful exploitation of these developments. Generally the loans cover 50 per cent of the necessary expenditure and in some circumstances can reach 80 per cent. Sometimes there is a requirement that the state share in the financial returns arising out of exploitation.

The Office for Industrial Promotion is an autonomous public body administered by representatives of the public and private sectors involved in R & D. It is under the trusteeship of the Ministry for

Economic Affairs and the Minister of Science Policy, and representatives of regional economic councils sit on its Board. The objectives of the office are as follows:

* to investigate, stimulate and promote innovations and encourage their development through to the final stage of industrialization and commercialization;
* to systematically investigate the possibilities of new lines of production and the expansion or diversification of existing activity;
* to encourage the undertaking of profitable projects in the private as well as the public economic sector either in partnership or alone;
* to promote the implementing of industrialization and commercialization of patents perfected through financial support from public funds;
* to study all economic industrial problems presented by the government, public financing bodies, regional economic councils and societies for regional economic development;
* to undertake any promotional action in favour of the national industrial sector.

Approaches for assistance can be made to the office by government, businesses, individual research workers and inventors, professional federations, research centres and university laboratories. These 'customers' pay for the services they receive but the office also receives a grant from the Ministry of Economic Development.

Co-ordination

Belgian science policy closely resembles the system defined here as concerted action. It has a Minister for Science Policy who is assisted by a powerful planning secretariat. There is a ministerial committee assisted by an interministerial commission of public servants and the main instrument of co-ordination and control is the science budget. All that distinguishes the Belgian system from a centralist one is the fact that the departments' R & D plans do not have to be in accordance with the principles laid down in a national plan.

In Belgium the new economic context has, in a sense, set the priorities that are evident in the research system. Attempts are being made to mobilize the whole R & D network in order to achieve national economic and social objectives. Thus, the concerted action programs have attempted to orient the universities more towards areas of social and economic concern and, like RANN and SUP in the United States, these attempts have met with limited success. National programs of research have also been put into operation in areas of great national priority which involve not only the government departments but the universities as well.

There is no evidence that national priorities in R & D are set in ways other than the consultative processes that have been set up. Inventories of the national R & D effort, compiled within the science policy minister's secretariat, serve as a background for the determination of priorities over almost the whole range of publicly-funded R & D. No formal mathematical techniques appear to be used in the decision-making processes.

CHAPTER 9

Science policy in the new economic context

During the 1960s, the Organization for Economic Co-operation and Development (OECD) played a key role in developing mechanisms at national level for formulating and implementing science and technology policies. Since that time it has maintained a continuing interest in science policy affairs through its Directorate for Science, Technology and Industry, and especially through the Committee for Scientific and Technological Policy of that Directorate. Periodically, expert committees are appointed to evaluate the impact that science and technology are having on the economic and social structures of member countries and to make recommendations on how new needs can be met and expectations fulfilled. Some of the reports of these expert groups have become landmarks in the development of the fundamental principles of science policy, and we have already referred to one of these, the Piganiol Report, issued in 1963, which coined the two terms 'policy for science' and 'science in policy' for two facets of science policy that had not previously been explicitly recognized.[1] The Piganiol group emphasized the need for OECD member countries to manage their scientific and technological resources as a whole and to understand the effects that new technological developments would have on areas of government responsibility apart from the purely scientific or technical. Following the report many European countries established science policy machineries that incorporated the OECD's preferred concerted-action model to greater or lesser degrees.

Towards the end of the 1960s expenditure on R & D began to level off or even decline in many European countries and North America as a more critical public attitude towards scientific and technological developments began to emerge. Though high levels of economic growth had been achieved, based largely on technological innovation, demands and expectations appeared to be changing in the industrialized nations. An OECD expert group, under the chairmanship of Harvey Brooks, was set up to identify the problems and trends that were beginning to emerge and to spell out the implications. The group concluded in its report, issued in 1971, that the slow-down in R & D expenditure was due partly to government efforts to control public expenditure and partly to the sceptical attitude that had developed towards all kinds of scientific and technological activity, especially towards the expensive civilian and military technological spectaculars. Adverse opinion focused especially on the following issues:

- the increasing sophistication of military hardware and the continuing threat of nuclear war;
- the general deterioration in the quality of life through the pursuit of economic growth fuelled by technological change;
- ethical and legal problems arising out of advances in the biological sciences;
- concentration on scientific and technological activity in the glamorous areas of high technology instead of in the non-glamorous service sector.

Though the Brooks Report considered further economic growth essential if living standards were to continue to rise, growth as measured by the market mechanism alone could no longer be regarded as a viable objective. Emphasis would have to shift from growth as a means of increasing personal incomes to growth as the means necessary for general social development. Those responsible for science policy would have to be more clearly involved in setting national goals, and the priorities amongst them, if these goals were to be realistic and if the R & D systems of the industrialized countries were to respond to them. But how was this to be done? Acknowledging that the achievement of a proper balance between centralist and pluralist policies was one of the most difficult and perplexing problems of science policy, the expert group could only suggest that OECD member countries should attempt to achieve a viable science policy that involved some blend of centralist and pluralist policies. Pluralism, the group believed, worked well in an expanding situation where goals are stable and unchanging, and the various sectors of government responsibility are well defined. The more centralized approach, generally adopted on mainland Europe, worked better where resources are limited and goals are relatively unstable and changing fairly rapidly.

The new tasks to be faced by the scientific community in the 1970s would necessarily be more complex than the old ones, having to take into account social, economic and cultural factors, and the need for co-ordination of R & D and innovation policy across the various sectors of government was recognized by the expert group. They were, however, unable to put forward any concrete suggestions as to how this co-ordination might be achieved. The classic government response to the need to co-ordinate activities across sectors was the establishment of interdepartmental committees or committees of the legislature and this was regarded by Brooks and his committee as ineffective. By implication, therefore, the Brooks report invited the science policy theorists and government decision-makers to go back to the drawing board so that viable mechanisms of interdepartmental co-ordination might be worked out.

There were three main recommendations of the Brooks Report. First, economic growth in the 1960s brought transformations that seemed to have rendered traditional concepts of economic policy obsolete and, since economic growth and technical progress were inextricably linked, the latter could no longer be regarded as an independent variable. Technology policies and socio-economic policies would have to be brought into a much closer relationship. Second, the group prophetically recognized that technical change would result in severe structural changes arising out of disparities between the different professions and occupations, between industrial sectors and between geographical regions. They recommended that the OECD and its member countries should undertake prospective studies of the likely impacts. Third, member countries were advised to channel their technological efforts into areas capable of producing alternative socially-oriented technologies, that is, technologies capable of contributing directly to the solution of infrastructural problems, to satisfying neglected social needs and to replacement of environmentally degrading techniques.[2]

The new economic context

In the twenty years that preceded the Brooks Report, there was rapid growth in the economies of virtually all of the OECD member countries. There was a drift of labour out of the agricultural into the manufacturing sector, technical advance was rapid, productivity increased at a steady pace and return on capital was relatively high. In the early to mid-1950s, the United States was ahead of all other industrialized nations in terms of GNP per worker. The United Kingdom, France and the Federal Republic of Germany came next with about half the GNP per worker of the United States, and Italy and Japan lagged behind these. By the beginning of the 1970s, however, the gap between the United States and the other countries had narrowed considerably. The major European countries and Japan had recovered from the effects of the Second World War and there was a new distribution of economic and industrial power. Though growth rates were beginning to falter, they were still satisfactory, employment was relatively full and prices were fairly stable. The subsequent decade saw a rapid deterioration in most of the OECD countries. In 1978 the OECD decided to establish an expert group to carry out a new assessment of the inter-action between scientific research, technological development and economic growth. This group, under the chairmanship of Bernard Delapalme, Director of Research of ELF-Aquitaine, set out to analyse the situation, to attempt to understand the roots of the problems be-setting the mature industrialized nations and to make recommenda-tions for action. The group report, which is considered to be a follow-up of the Brooks Report, was presented to the OECD in 1979 and sub-sequently published under the title *Technical Change and Economic Policy*.[3] The Delapalme Report said that the world had changed dramatically and irreversibly in the ten years since the publication of the Brooks Report for the following reasons. First, the unemployment that was beginning to appear in the industrialized countries at the beginning of the 1970s increased to high levels after the 1973 oil crisis and seemed not to respond to the application of conventional economic policy instruments. The relationship of prices and levels of employment to overall economic activity, and thus to each other, appeared to be less predictable; even with high unemployment and spare industrial capacity there were few reductions in prices or wages. The uncertainty of this situation had spilled over into other areas of economic activity, especially investment, and was creating a difficult environment for economic recovery. Second, the newly industrializing countries were playing an increasingly important role in international trade. The Asian region, in particular, has emerged as one of the toughest competitors in international markets. Writing about South-East Asian performance in the 1970s, Wyllie said:

> In this period few other regions of the world have so consistently notched up high growth rates, growth which is measured in double digits, a term that most Western businessmen are perhaps more familiar with when re-ferring to inflation.[4]

Since 1975 Hong Kong has had an average annual growth rate of 11 per cent, Singapore 8.5 per cent and Malaysia 8 per cent. Countries that started the 1970s as less developed are now described as newly industrializing, and by the end of the 1980s may have joined the ranks of the developed countries. This growth cannot be explained simply as satisfying domestic consumer and industrial demand. In 1980 Hong Kong was sixteenth in the list of the world's top trading nations and amongst the world's leaders in the export of garments, toys, electronic goods, watches and binoculars. The region's increasingly sophisticated technological base has been a crucial factor in the success of the Asian countries' entry to world markets. There is a willingness to adopt new technology and a recognition by both government and industry of the importance of international competitiveness. In effect the 'Japanese miracle' is being re-enacted in these newly industrializing nations and the impact on Western economies will be quite severe in the 1980s and 1990s. The Delapalme group considered it essential for OECD countries to change their products to replace those that the less developed and newly industrializing countries are now making for themselves.

The costs of production have changed during the decade as a result of massive increases in the price of crude oil. The industrial structure of OECD countries, based on pre-1973 costs, had become ill-adjusted if not completely obsolete in the new context and a restructuring of the existing industrial apparatus was becoming essential. Should such structural adjustment not take place the economies would continue to stagnate.

In the last decade public expectations of education, health care, and other public services and cultural activities have altered. In many countries, spending on consumer goods has not increased as fast as overall expenditure on services, and the proportion of the workforce employed in the service sector has rapidly increased. If demand for consumer goods continues to fall then it will become increasingly difficult to support the new jobs created in the expanding service sector.

The trend identified in the Brooks Report of greater public attachment to protection of the environment has continued. Resources like air and water that had at one time been regarded as 'free' were now given value, and costs that would previously not have been incurred are now a charge on manufacturers, developers and ultimately the public.

Changes have occurred in the R & D systems of OECD countries. In the manufacturing sector there has been a shift in emphasis towards short-term, safe projects in response either to the unfavourable economic climate or as a result of changed corporate behaviour and, in general, there has been a slow down in the growth of industrial R & D.

During the decade there was a decline in the growth of labour productivity and of employment in the manufacturing sector of the leading industrial nations. This trend in employment in the manufacturing sector was evident before the 1973 oil crisis which contributed to its subsequent acceleration. Other reasons for the decline in productivity growth and employment levels in the sector have already been mentioned — entry of the newly industrializing nations to international markets, for example, and inflation. In addition, the establishment and enforcement of new environmental and safety regulations have meant that a great deal of new investment is going into equipment imposed by the regulations and this does nothing for labour productivity. The restrictive fiscal and monetary policies that have been pursued by many governments, especially in Europe, have contributed to economic stagnation. These policies have been inspired by a desire to contain public expenditure, but the atmosphere of restraint rather than encouragement of demand has had a restricting effect on

investment in new plant and machinery. The rate at which new technology is brought into play has been slowed, and willingness to carry
out R & D has declined.

The trend towards low productivity growth and employment levels
in the manufacturing sector has not, however, been universal. The
electronics industry, in particular, has shown no noticeable decrease in
the rate of innovation or growth in productivity. But it is this industry
that has precipitated one of the more serious problems facing governments in the 1980s — the prospect of large-scale structural unemployment through technological change.

Technological change and unemployment

According to the Delapalme Report, the high levels of unemployment
in the OECD countries are related to their sustained high rates of
inflation; if demand for goods and services could be stimulated by tax
cuts and by loosening the constraints on money supply unemployment
levels would fall. But there are fears that stimulation of demand in this
way would increase inflation even further and exacerbate balance of
payments and exchange rate problems. The rate at which investment
can proceed without returns diminishing depends on the flow of technical innovations. Rapid innovation stimulates productivity and also
stimulates demand for goods and services by encouraging investment.
The ability of domestic suppliers to compete with foreign imports is
enhanced in a period of rapid innovation, which eases balance of
payments problems, and governments can increase the money supply
to keep demand up. Countries with poor innovative capacity, low investment and productivity cannot compete and cannot absorb wage
increases without raising prices. Unemployment rises and their governments find it difficult and costly to play the strong stimulating role that
is necessary under these conditions.

But when a technical advance introduces new techniques that make
it cheaper to produce a unit of output with more capital and less
labour, businesses are encouraged to shed labour and increase capital
expenditure. This is occurring in many important industries and the
reduction of employment in the manufacturing sector partly reflects
this. The service industries have absorbed some of the fall-out from
manufacturing but employment in this sector, too, has been affected
by technical change. The impact of microelectronic technology in
particular has been a source of great concern and a number of major
studies have been carried out to try to determine its effect on employment. The results of these studies have not been conclusive. The
United Kingdom Central Policy Review Staff (CPRS) had this to say:

There is great public interest in the likely effects of microprocessors on our way of life but even more unease at the possible effects on employment. In the work so far we have been struck by the contrast between the vehemence of those who claim that microelectronics will have a catastrophic effect on employment and the inadequacy of the analysis underlying the certainty of that prediction.[5]

The CPRS itself conducted a survey of the impact of microelectronics on employment in the United Kingdom Civil Service and in a number of industries. The conclusion was reached that in the Civil Service there had been an increase of at least 30,000 jobs in the period reviewed in areas that one would have expected to be affected by computer installation. In some cases, the computer applications themselves suggested new areas of work and new services that the government could develop with benefit to the community. Though the overall effect of the introduction of word processing and other computer applications to the Civil Service had been to restrain the growth of clerical employment, it certainly did not reduce the overall levels. The CPRS was unable to generalize about the introduction of microprocessor technology in the industries that they studied (printing, textiles, food processing and car manufacture). The position they adopted towards microelectronic technology was, however, quite clear:

> Unless the United Kingdom does all it can to accelerate its own microprocessor revolution it will be subjected to even fiercer international competition in the home market and fail to win overseas markets. The United Kingdom must be ahead of its competitors in productivity increases which the microprocessors make possible.

The view of the CPRS was that without the vigorous pursuit of labour-saving capital-intensive technology which utilized the latest developments in microelectronic technology Britain's economic condition could only worsen. They argued that the fears frequently expressed about the effects of such technology on employment do not adequately reflect the employment-creating potential of microelectronics.

Subsequent to the CPRS report, a study of the implications of microelectronic technology was commissioned by the United Kingdom Department of Employment. The Sleigh Report which resulted concluded that

- the decline in employment in the manufacturing sector in Britain was not inevitable, nor was it a symptom of de-industrialization;
- growth of employment in the service sector was not necessarily associated with losses in the manufacturing sector in the long term;
- whereas the overall employment effect of microprocessors was virtually impossible to gauge, empirical evidence suggested that, as

in the past, technical change had been beneficial to output and employment;

- in the service sector job losses in the less skilled clerical and sub-clerical areas would occur in the next 5-10 years but growth of new and existing services was likely to result in offsetting job opportunities.[6]

The Sleigh group suggested that changes to British industry arising out of the introduction of microelectronic technology would be of two kinds. Some innovations would enable production to be carried out more continuously and others would substitute machines for people without necessarily changing the characteristics of the production process. The group considered that the former changes were likely to have a greater effect on employment but that the overall effect would depend on whether new technology was used as a means of expanding output or simply as a means of cutting labour and other production costs. In many areas of industry that were already highly automated the main effect of new technology would be to introduce more sophisticated control systems with negligible effects on overall manpower levels.

The International Labour Office (ILO) in Geneva undertook an extensive study of the impact of microelectronics in 1980 and came to the conclusion that from the existing figures on job creation and job displacement, substantial losses have taken place and the rate of job loss resulting from microelectronic technology will accelerate in the 1980s. According to the ILO report, European society was undergoing a transition from one in which unemployment was the result of fluctuating economic circumstances to one in which the full potential workforce would no longer be needed to produce the array of goods and services. Shorter working hours, early retirement and other measures designed to spread the available work over a greater number of people would not, according to this study, alleviate the problem.[7]

In 1979, too, in The Netherlands, an advisory group on the social effects of microelectronics was formed by the then Minister for Science Policy and asked to prepare a report. The group's report said that if present policies persisted in The Netherlands there would be an unacceptable shortage of employment by the 1990s. Though they did not take the pessimistic view that work as we know it would disappear altogether allowance would have to be made for reduction in the opportunities for work in the current sense of the word. Like the British reports, the Dutch advisory group advocated continued expansion and intensification of national expertise in microelectronics. Its specific recommendations were for the establishment for a centre for microelectronics and the establishment of a technology assessment facility to monitor the effects of technological change.[8]

In Australia, a Committee of Inquiry into Technological Change

(CITCA) was established following a major dispute concerning the introduction of new technology by Telecom Australia. Much has been written about the four volume report issued by the committee and it is beyond the scope of this book to enter into the debate. Stubbs has said that given the circumstances under which the committee was set up, 'any report which was anti-technology would have been as absurd as it was unlikely'[9] and this applies to the other government-sponsored reports considered here as well. The committee concluded that though technological changes have the capacity to reduce the number of jobs required to produce a given level of output, change will occur with least disruption in those companies who keep abreast of new technological developments. As will be seen by the following quotations its thinking paralleled that of the Sleigh Report:

> . . . the pressures of international competition are such that the employment consequences of not adopting cost-saving technological changes may be far greater in the long run than the adjustment required in the first round if change is accepted. . . . 'jobless growth' will, of course, occur and has been occurring for centuries in particular firms or industries or even in sectors of the economy. To postulate 'jobless growth' in the economy as a whole, however, is to overlook the substantial unmet needs and wants in the community or to believe that the income generated by technological change will not be sufficient or will not be distributed or spent in such a way as to create sufficient employment in the process of satisfying those needs.[10]

As a result of the CITCA report on technological change, a standing committee of ASTEC, the Technological Change Committee, was established to monitor the effects of technical change in Australia.

Science policy trends

In general, science policies are being shaped by the need to adjust to the new social and economic context brought on first by the oil crisis, then by the entry of the newly industrializing nations to the international economic arena, and finally by the employment effects of new technology. Obviously, the severity of these problems varies from nation to nation. Some have been cushioned from the effects of oil price rises while others are totally dependent on oil imports. Some are rich in coal and other natural resources, others are not. There is also no uniformity in the effects that the newly developing nations have had on domestic and export markets. Defence expenditure is a greater burden in the United States, the United Kingdom and Japan than it is in some of the other countries reviewed. The varying degrees to which the different problems are experienced in the countries in question have determined the thrust of their science policies. As the OECD has said, its members have entered the 1980's with different problems, expectations and priorities, and different strategies for dealing with them. [11]

A theme running through the strategies of almost all of the OECD countries is the need to stimulate industrial R & D. Studies of the effects of microelectronic technology have only served to increase determination to acquire rapidly innovating manufacturing sectors and it is universally accepted that government intervention is necessary in order to achieve that aim. Some countries have decided that intervention across the board is unrealistic and relatively ineffective and have pursued a key industry or key technology policy in which selective support is concentrated in industries within the chosen area or technology. Others have steadfastly refused to adopt such a policy, being unconvinced of its value in the light of spectacular failures like Concorde.

There has been a change in attitude towards R & D expenditure which, for many years, has been almost static in real terms or has even declined. Reflecting their commitment to rapid innovation, and despite economic policies that are attempting to hold down public expenditure, a number of countries have set targets for their R & D expenditure as a percentage of GNP. Japan intended to have a GNERD of 2.5 per cent of the GNP by 1985 rising to 3 per cent by 1990; Canada has, as we have seen, set a target of 1.5 per cent by 1985 and in the United States there is a long-standing commitment to raise the level of government expenditure on basic research.

There is now a greater awareness of the need to have strategic plans and to co-ordinate R & D efforts, and science budgets are becoming an increasingly important feature of national science policies, especially in those countries that have tried to adopt a concerted action science policy system. Even in decentralized systems forward plans are required from the various elements of the system. In the United Kingdom, for example, the Advisory Board for the Research Councils requires each Council to provide an indicated financial plan for the two years subsequent to the expenditure year, and in the Federal Republic of Germany the BMFT major programs give forward estimates of up to three years. In the United States the Five Year Outlook is an attempt to identify emerging national problems to which R & D might be expected to contribute.

In order to tie R & D programs more closely to national needs, especially in areas that cross departmental boundaries (energy, transportation, communications, marine science, space activities, for example), there has been an increased emphasis on co-ordination, and co-ordinating bodies have been established to monitor the areas under control. Thus, in Canada, there are national programs controlled by interdepartmental committees, co-ordinated by a lead department, which provides a secretariat. The lead departments are required to report to Cabinet and Treasury Board on the utilization of resources and changing of priorities when such changes are desirable. In Belgium, the national research programs, controlled by the Ministry of Science Policy and started with pump-priming funds from the Prime Minister, are an attempt to acquire a better scientific and techno-logical understanding of the problems that are of interest to a number of departments which go beyond the limits of departmental research budgets. In The Netherlands there are national programs of research co-ordinated by steering groups in which the Ministry of Education and Science is represented, and sector councils have been established to facilitate the co-ordination and organization of research in accor-dance with social priority areas.[12] Finally, in the Federal Republic of Germany, the BMFT co-ordinates its major national programs of research in which the programs, sub-programs and projects are published and tenders called to which the various performance sectors can respond.

The existence of these national programs and co-ordinating bodies shows a serious intention by the countries concerned to employ their scientists and technologists in a concerted manner to attack the problems that have emerged in the new economic context. Govern-ments seem no longer content to adopt Alexander Strange's dictum that 'You cannot pay a man to discover but you can put him in a position favourable to discovery.'[13] In the new social and economic context each of these governments is attempting to place its scientists in a position favourable to making discoveries that will reflect govern-ment priorities and be of benefit to the nation as a whole.

References

Chapter 1

[1] E. Shils, 'Introduction', in E. Shils, editor, *Criteria for Scientific Development* (Cambridge, Mass., 1968), p. ix

[2] S.A. Lakoff, 'The Third Culture', in S.A. Lakoff, editor, *Knowledge and Power* (New York, 1966), p. 14

[3] S. Toulmin, 'The Complexity of Scientific Choice', *Minerva*, IV (1966), pp. 155-69

[4] A. Condorcet, *Sketch for a Historical Picture of the Progress of the Human Mind*, edited by Garat-Cabanis (8 vols, Paris, 1804), VIII, p. 598

[5] J-J. Salomon, *Science and Politics* (London, 1973), p. 15

[6] ibid., pp. 17-18

[7] C. Babbage, *Reflections on the Decline of Science in England and on Some of Its Causes* (London, 1830), p. 1

[8] ibid., p. 19

[9] J.G. Crowther, *Statesmen of Science* (London, 1965), p. 241

[10] ibid., p. 254

[11] R.M. MacLeod, 'The Support of Victorian Science — The Endowment of the Research Movement in Great Britain, 1868-1900', *Minerva*, IX (1977), pp. 197-230

[12] R. Moseley, 'The Origins and Early Years of the National Physical Laboratory: A Chapter in the Pre-History of British Science Policy', *Minerva*, XVI (1978), pp. 222-50

[13] ibid., pp. 223-4

[14] J.B. Poole and K. Andrews, editors, *The Government of Science in Britain* (London, 1968), p. 55

[15] D.S. Greenberg, *The Politics of Pure Science* (New York, 1971), p. 58

[16] H. Sapolsky, 'Science, Technology and Military Policy', in I. Spiegel-Rösing and D. de Solla Price, editors, *Science, Technology and Society* (London, 1977), p. 445

[17] Poole and Andrews, op. cit., p. 62

[18] Viscount Haldane of Cloan, *Report of the Machinery of Government Committee* (London, 1918), p. 66

[19] H. Rose and S. Rose, *Science and Society* (Harmondsworth, 1971), p. 73

[20] W.R. Brode, 'Development of a Science Policy', in R.H. Kargon, editor, *The Maturing of American Science* (Washington, 1974), p. 159

[21] D. Kevles, *The Physicists* (New York, 1978), p. 105

[22] V. Bush, *Science, the Endless Frontier* (Washington, 1945), pp. 3, 4

[23] ibid., p. 12

[24] ibid., p. 19

[25] J.L. Penick, C.W. Pursell, M.G. Sherwood and D.C. Swain, *The Politics of American Science* (Cambridge, Mass., 1965), p. 22

[26] ibid., p. 23

[27] W.H. Lambright, *Governing Science and Technology* (New York, 1976), p. 195
[28] D.K. Price, *The Scientific Estate* (New York, 1968), pp. 239-40
[29] Penick et al., op. cit., pp. 231, 232
[30] E.G. Mesthene, editor, *Ministers Talk About Science* (Paris, 1965), p. 160

Chapter 2

[1] V. Bush, *Science, the Endless Frontier* (Washington, 1945), p. 12
[2] ibid., p. 19
[3] T.H. Huxley, 'Science and Culture', in *Science, Culture and Other Essays* (New York, 1882), p. 26
[4] Australian Science and Technology Council, *Basic Research and National Objectives* (Canberra, 1981), p. 3
[5] E. Mansfield, *Economics of Technological Change* (New York, 1968), pp. 10-11
[6] C. Freeman, 'Technical Change and Unemployment', in S. Encel and J. Ronayne, editors, *Science, Technology and Public Policy: An International Perspective* (Sydney, 1979), pp. 53-76; S.C. Hill, 'Ideas Whose Time Has Come: The Waves of Social Consequence from Scientific and Technological Advance', in A.T.A. Healy, editor, *Science and Technology for What Purpose? An Australian Perspective* (Canberra, 1979), pp. 125-42
[7] E.F. Denison, *The Sources of Economic Growth in the United States and the Alternatives Before Us* (New York, 1962), pp. 229-55
[8] Z. Grilliches, 'Research Costs and Social Returns: Hybrid Corn and Related Innovations', *Journal of Political Economy*, LXVI (1958), pp. 419-31.
[9] J. Schmookler, *Invention and Economic Growth* (Cambridge, Mass., 1966), pp. 104-15 and 204-7
[10] ibid., p. 207
[11] R.L. Nelson, 'The Simple Economics of Basic Scientific Research', *Journal of Political Economy*, LXVII (1959), pp. 297-306
[12] K.J. Arrow, 'Economic Welfare and the Allocation of Resources for Invention', in National Bureau of Economic Research, *The Rate and Direction of Inventive Activity* (Princeton, 1962), pp. 609-25
[13] K. Pavitt and W. Walker, 'Government Policies Towards Industrial Innovation: A Review', *Research Policy*, V (1976), pp. 15-96
[14] *Economic Report of the President* (Washington, 1972), p. 126
[15] National Science Foundation, *Science and Technology Annual Report to Congress* (Washington, 1980), p. 55
[16] H.F. Cruise, 'Science, Technology and Government: The Relevance of Government Assistance', *Paper delivered to Second Division Seminar* (Canberra, 1980), p. 1
[17] R. Maxwell-Hyslop, 'But how many will fly in it?', *The Economist*, 241 (1971), pp. 85-86
[18] Cruise, op. cit., p. 5
[19] N. Wade, 'Why Government Should Not Fund Science', *Science*, CCX (1980), p. 33
[20] Council for Scientific Policy, *Second Report on Science Policy* (London, 1967), p. 12
[21] H.G. Stever, 'Introductory Remarks', in G. Strasser and E.M. Simmons, editors, *Science and Technology Policies* (Cambridge, Mass., 1973), p. xx
[22] J.P. Wild, 'Interscan', in Science and Industry Forum, *From Stump-Jump Plough to Interscan: A Review of Invention and Innovation in Australia* (Canberra, 1977), p. 68
[23] C. Cherry, M. Gibbons and J. Ronayne, 'The Origins of the Air Turbine Dental Handpiece', *British Dental Journal*, CXXXVI (1974), pp. 469-72
[24] J. Jewkes, D. Sawers and R. Stillerman, *The Sources of Invention* (London, 1958)
[25] ibid., p. 224

[26] J. Jewkes, D. Sawers and R. Stillerman, *The Sources of Invention*, second edition (London, 1969), pp. 225-6

[27] C. Freeman, *The Economics of Industrial Innovation* (Harmondsworth, 1974), pp. 69-73

[28] B.R. Williams, 'Research and Economic Growth — What Should We Expect?', *Minerva*, III (1964), pp. 57-71

[29] D.J. de Solla Price, 'Is Technology Independent of Science? A Study in Statistical Historiography', *Technology and Culture*, VI (1965), p. 568

[30] K. Kreilkamp, 'Hindsight and the Real World of Science Policy', *Science Studies*, 1 (1971), p. 44

[31] IIT Research Institute, *Technology in Retrospect and Critical Events in Science* (Washington, 1968)

[32] Science Policy Research Unit, University of Sussex, *Success and Failure in Industrial Innovation* (London, 1972), pp. 5-6

[33] J. Langrish, M. Gibbons, W.G. Evans and F.R. Jevons, *Wealth from Knowledge* (London, 1972)

[34] J. Langrish, 'Technology Transfer: Some British Data', *Research and Development Management*, I (1971), p. 135

[35] G. Pragier and J. Ronayne, 'A Criticism of the Use of Citation Analysis in Studying the Science-Technology Relationship', *Journal of Chemical Information and Computer Sciences*, XV (1965), pp. 155-7

[36] Council for Scientific Policy, *Third Report on Science Policy* (London, 1972), p. 32

[37] Kreilkamp, op. cit., pp. 63-6

[38] E. Layton, 'Conditions of Technological Development', in I. Spiegel Rösing and D. de Solla Price, editors, *Science, Technology and Society* (London, 1977), p. 207

[39] Langrish et al., op. cit., p. 120

[40] E.B. Chain, *Proceedings of the Royal Society (B)*, CLXXIX (1971), p. 301

[41] G. Pragier, *Pharmaceutical Innovation: Three Case Studies of Corporate Achievement Meriting the Queen's Award to Industry* (Manchester, 1974), p. 81

[42] M. Gibbons and R. Johnston, 'The Roles of Science in Technological Innovation', *Research Policy*, III (1974), pp. 220-42

[43] ibid., p. 241

[44] Freeman, op. cit., p. 194

[45] K.G. Gannicott, 'Research and Development Incentives', in Committee of Inquiry into Technological Change in Australia, *Technological Change in Australia*, Vol. 4 (Canberra, 1980), pp. 287-314

[46] Office of the Director of Defense Research and Engineering, *Project Hindsight*, (Washington, 1967), p. 96

[47] Schmookler, op. cit., p. 256

[48] See, for example, S.C. Gilfillan, *The Sociology of Invention* (Chicago, 1935) and S. Hollander, *The Sources of Increased Efficiency* (Cambridge, Mass., 1965)

[49] Hollander, op. cit., p. 186-7

[50] D. Mowery and N. Rosenberg, 'The Influence of Market Demand upon Innovation: A Critical Review of Some Recent Empirical Studies', *Research Policy*, VIII (1979), pp. 102-53

[51] S. Myers and D.G. Marquis, *Successful Industrial Innovations. A Study of Factors Underlying Innovation in Selected Firms* (Washington, 1979)

[52] Mowery and Rosenberg, op. cit., p. 107

Chapter 3

[1] R.A. Rettig, *Criteria for the Allocation of Resources to Research and Development* (Ohio, 1971), p. 9

[2] National Institutes of Health, *Issue Paper: What Should be the Level of Support for*

Biomedical Research During the Next Five Years? (Washington, 1971), p. 10

[3] H. Brooks, 'The Problem of Research Priorities', *Daedalus*, CVII (1978), p. 172

[4] A.M. Weinberg, 'Criteria for Scientific Choice', *Minerva*, I (1963), p. 161

[5] OECD, *Science, Growth and Society* (Paris, 1971), p. 16

[6] A.M. Weinberg, 'Criteria for Scientific Choice II', *Minerva*, III (1964), p. 12

[7] C. Kaysen, 'Federal Support of Basic Science', in National Academy of Sciences, *Basic Research and National Goals* (Washington, 1965), p. 139

[8] Lord Rothschild, 'The Organisation and Management of Government R and D', in Lord Privy Seal, *A Framework for Government Research and Development* (London, 1971), p. 7

[9] Weinberg, op. cit., p. 14

[10] H. G. Johnson, 'Federal Support of Basic Research: Some Economic Issues', in National Academy of Sciences, op. cit., pp. 132-3

[11] Brooks, op. cit., p. 177

[12] A.V. Hill, *The Ethical Dilemma of Science* (New York, 1960), pp. 206-8

[13] C.H. Waddington, *The Scientific Attitude* (Harmondsworth, 1941), p. 58

[14] J.R. Baker, *The Scientific Life* (London, 1942), pp. 48-9

[15] M. Polanyi, 'The Republic of Science: Its Political and Economic Theory', *Minerva*, I (1962), p. 55

[16] ibid., p. 62

[17] Brooks, op. cit., p. 177

[18] Senate Special Committee on Science Policy, *A Science Policy for Canada* (4 vols., Ottawa, 1970-6), I, pp. 269-72

[19] M. Polanyi, 'Science: Academic and Industrial', *Journal of the Institute of Metals* (LXXXIX, 1960-1), p. 404

[20] G. Herzberg, 'Bureaucracy and the Republic of Science', *Impact of Science on Society*, XXII (1972), pp. 105-10

[21] ibid., p. 107

[22] J.D. Babbitt, 'Science Policy as Ideology: The Collectivization and Socialization of Research', *Canadian Research and Development*, IX (1976), p. 22

[23] J.R. Philip, 'Towards Diversity and Adaptability: An Australian View of Governmentally Supported Science', *Minerva*, XVI (1978), pp. 397-415

[24] R.K. Merton, 'Science and Technology in a Democratic Order', *Journal of Legal and Political Science*, I (1942), p. 118

[25] ibid., p. 402

[26] C. Carter, 'The Distribution of Scientific Effort', *Minerva*, I (1963), p. 180

[27] C. Kaysen, op. cit., p. 160

[28] H. Brooks, op. cit., p. 178

[29] J. Irvine and B. Martin, 'Assessing Basic Research: The Case of the Isaac Newton Telescope', *Social Studies of Science*, XIII (1983), pp. 49-86; B. Martin and J. Irvine, 'Assessing Basic Research: Some Partial Indicators of Scientific Progress in Radio Astronomy', Research Policy XII (1983), in the press

[30] M. Gibbons, 'The CERN 300 GeV Accelerator: A Case-Study in the Application of Weinberg's Criteria', *Minerva*, VIII (1969), pp. 180-91

[31] H. Brooks, op. cit., p. 176

[32] J-J. Salomon, 'Institutional Aspects', in OECD, *Problems of Science Policy* (Paris, 1967), p. 41

[33] Senate Special Committee on Science Policy, *A Science Policy for Canada* III, pp. 629-43 and W. Hutter, 'Science Policy in The Netherlands', *Planning and Development in The Netherlands*, VIII (1972), pp. 107-8.

[34] R. Brickman, 'Comparative Approaches to R and D Policy Co-ordination', *Policy Sciences*, XI (1979), pp. 73-91

[35] J. Schmandt, 'Science Policy: One Step Forward, Two Steps Back', in J. Haberer, editor, *Science and Technology Policy* (Lexington, 1977), pp. 14-15

[36] R. Brickman and A. Rip, 'Science Policy Advisory Councils in France, The Netherlands and the United States, 1957-77: A Comparative Analysis', *Social Studies of Science*, IX (1979), pp. 167-98

[37] C.E. Lindblom, 'The Science of Muddling Through', *Public Administration Review*, XIX (1959), pp. 79-88

[38] Y. Dror, 'Muddling Through — Science of Inertia?', in A. Etzioni, editor, *Readings on Modern Organizations* (Englewood Cliffs, 1969), pp. 166-71

[39] K. Pavitt, 'Analytical Techniques in Government Science Policy', *Futures*, IV (1972), pp. 5-12

[40] ibid., p. 10

[41] H. Brooks, personal communication

[42] E.M. Grabb and D.L. Pyke, 'An Evaluation of the Forecasting of Information Processing Technology and Applications', *Technological Forecasting and Social Change*, IV (1973), pp. 143-50

[43] S. Encel, P.K. Marstrand and W. Page, *The Art of Anticipation* (London, 1975), p. 79

[44] ibid., p. 120

[45] ibid., p. 124

[46] Harvey Sapolsky, *The Polaris Development* (Cambridge, Mass., 1972)

Chapter 4

[1] A.H. Teich, 'The Development of Science Planning in the United States: From Endless Frontier to the Five Year Outlook', Paper delivered to the *Fourth International Conference on the Management of Research, Development and Education* (Wroclaw, 1980), p. 4

[2] J. Carter, *Message to Congress on Science and Technology* (Washington, 1979)

[3] W. Carey, 'Foreword', in W.H. Shapley, *Research and Development in the Federal Budget FY 1977* (Washington, 1977), p. v

[4] *The National Science and Technology Policy, Organization and Priorities Act* 1976

[5] Teich, op. cit., pp. 11-12

[6] ibid., p. 13

[7] P. Handler, 'On Reports', in *The National Research Council in 1978* (Washington, 1978), p. 29

[8] National Academy of Sciences, *Basic Research and National Goals* (Washington, 1965), pp. 22-4

[9] W.H. Shapley, A.H. Teich, G.J. Breslow and C.V. Kidd, *Research and Development AAAS Report VI* (Washington, 1981), pp. 3-4

[10] ibid., p. 11

[11] ibid., pp. 28-9

[12] ibid., p. 70

[13] National Science Foundation, 'Criteria for the Selection of Research Projects', in *NSF Guide to Programs 1977* (Washington, 1977), p. viii

[14] L.V. Blankenship and W.H. Lambright, *University Research Centers: A Comparison of the NASA and RANN Experiences* (Washington, 1977), p. 186

[15] ibid., p. 185

[16] S. Schneyer, Division of Program Analysis, Office of the Director, National Institutes of Health: interview with author

[17] National Institutes of Health, *Issue Paper: What Should Be the Level of Support for Biomedical Research During the Next Five Years?* (Washington, 1971), pp. 15-25

[18] S. Schneyer: interview with author

[19] B. Bozeman and L.V. Blankenship, 'Science Information and Governmental Decision-Making: The Case of the National Science Foundation', *Public Administration Review*, XXXIX (1979), p. 55

Chapter 5

[1] N. Vig, 'Policies for Science and Technology in Great Britain: Post-war Development and Reassessment', in T.D. Long and C. Wright, editors, *Science Policies of Industrial Nations* (New York, 1975), p. 65

[2] Lord Rothschild, 'The Organization and Management of Government R and D', in Lord Privy Seal, *A Framework for Government Research and Development* (London, 1971), pp. 1-25

[3] P. Gummett, *Scientists in Whitehall* (Manchester, 1980), pp. 165-67

[4] T.M. Copestake and G.M. White, 'Priorities for Applied R and D: Some Real Problems and Practical Solutions', Paper presented to *OECD Forum Discussion of Priorities in R and D Policy* (Paris, 1981), p. 8

Chapter 6

[1] R.D. Johnston, *Key Issues for Australian Science Policies* (Wollongong, 1980), p. 18

[2] ASTEC, *Industrial Innovation: A Discussion Paper* (Canberra, 1979) and ASTEC, *Industrial Research and Development: Proposals for Additional Incentives* (Canberra, 1980)

[3] Study Group on Structural Adjustment, *Report* (2 vols., Canberra, 1979), 1, pp. 7.34-37.

[4] *Science in Australia: Proceedings of a Seminar Organized by the Australian National University* (Melbourne, 1951)

[5] Commonwealth Parliamentary Debates (Canberra, 1965), p. 274

[6] Australian Labor Party, *Platform Constitution and Rules* (Canberra, 1965), p. 15

[7] Science and Industry Forum, *Government Approaches to Science: An Address to the Forum by the Minister for Education and Science* (Canberra, 1969), p. 12

[8] P. Pockley, 'Under a Cloud of Committees', *Nature*, CCLXVII (1977), p. 476

[9] E.G. Whitlam, 'A National Science Policy', *Search*, I (1970), p. 135

[10] S. Encel, *Pushing the Barrow Uphill: The Establishment and Re-establishment of ASTEC* (University of New South Wales, 1981), pp. 11-12

[11] OECD, *Examiners' Report: Australia* (Canberra, 1974)

[12] M. Fraser, in House of Representatives, *Hansard*, 21 March 1974, p. 2

[13] J. Ronayne, 'Further Thoughts on Diversity and Adaptability in Australian Science Policy', *Minerva*, XVIII (1979), pp. 445-58

[14] Science Task Force, *Report to The Royal Commission on Australian Government Administration* (Canberra, 1975), p. 6

[15] ibid., p. 21

[16] ASTEC, *The Bureau of Mineral Resources, Geology and Geophysics (BMR): A Report to the Prime Minister* (Canberra, 1978)

[17] *Report of the Committee of Inquiry into the Bureau of Meteorology* (Canberra, 1977)

[18] *Internal Review into Objectives and Procedures of the Defence Science and Technology Organisation* (Canberra, 1980); *Independent External Review of the Defence Science and Technology Organisation* (Canberra, 1980)

[19] Encel, op. cit., p. 12

[20] B. Nelson, 'Australia: Education and Science Are Looking Up Down Under', *Science*, CLX (1968), p. 173

[21] Science Task Force, op. cit., p. 48

[22] G. McAlpine and R. Badger, *Bases for Science and Technology Policy* (Canberra, 1981)

[23] ibid., p. 2

[24] ibid., p. 34

[25] G. Currie and J. Graham, *The Origins of CSIRO* (Melbourne, 1966), p. 191

[26] D.P. Mellor, 'The Role of Science and Industry' in *Official History of Australia in the War of 1939-45*, (5 vols., Canberra, 1957), V, p. 23

[27] J. Ronayne, 'Scientific Research, Science Policy and Social Studies of Science and Technology in Australia', *Social Studies of Science*, VIII (1978), p. 370

[28] G. Currie and J. Graham, 'CSIR 1926-1939', *Public Administration* (Sydney), XXXIII (1974), pp. 232-3

[29] Sir Frederick White, 'CSIR to CSIRO — The Events of 1948-1949', *Public Administration* (Sydney), XXXIV (1975), p. 281

[30] D.P. Mellor, op. cit., p. 23

[31] G. Currie and J. Graham, op. cit., p. 236

[32] S.H. Bastow, 'CSIRO and the Universities', *Vestes*, IV (1961), p. 21

[33] ibid., p. 23

[34] *Report of the Committee on Australian Universities* (Canberra, 1957)

[35] OECD, op. cit., p. 15

[36] CSIRO Advisory Council, *Report of the Committee on the Relationship between CSIRO and the Universities* (Canberra, 1967)

[37] ASTEC, *Interaction Between Industry, Higher Education and Government Laboratories: A Report to the Prime Minister* (Canberra, 1980), p. 6

[38] G. Julius, *Improvement in Industry*, Transcript of Speech to Victoria League, Melbourne, June 1929, p. 6

[39] Y. Esplin, *Technology Transfer Takes Two*, Paper delivered to Section 37, ANZAAS, May, 1982, p. 3

[40] Sir David Rivett, *The Application of Science to Industry in Australia* (Brisbane, 1944), p. 2

[41] S.H. Bastow, 'Research in Manufacturing Industry in Australia', *Journal of the Institution of Engineers in Australia*, XXXVI (1964), pp. 37-8

[42] *Report of the Independent Inquiry into the Commonwealth Scientific and Industrial Research Organization* (Canberra, 1977), p. 130

[43] The Parliament of the Commonwealth of Australia, *A Bill for an Act to Amend the Science and Industry Research Act 1949* (Canberra, 1978), p. 3

[44] CSIRO Annual Report 1980/81 (Canberra, 1981), pp. 37-8

[45] UNESCO, *Method for Priority Determination in Science and Technology* (Paris, 1978)

[46] ASTEC, *Internal Report on the ASTEC Workshop on National Objectives and Research Priorities* (Canberra, 1982), p. 8

Chapter 7

[1] *Sixth Report of the Federal Government on Research* (Bonn, 1980), p. 10

[2] *Fifth Report of the Federal Government on Research* (Bonn, 1976), p. 16

[3] ibid., p. 8

[4] ibid., p. 15

[5] Senate Special Committee on Science Policy, *A Science Policy for Canada* (Ottawa, 1970), p. 128

[6] H. Faulkner, House of Commons Debates, Second Session, Thirtieth Parliament (Ottawa, 1977), p. 4668

[7] *Guide to Policy and Expenditure Management System* (Ottawa, 1980), p. 1

[8] Ministry of State for Science and Technology, *Strategic Overview 1983/1984-1985/86* (Ottawa, 1982), p. 9

[9] Anon., 'Johnston Appointment Gives Science a Central Role', *Science Notes*, I (1982), p. 1

[10] Anon., 'GERD Reaches All-Time High', *Science Notes*, 2 (1983), p. 1

[11] Science Council of Canada, *Annual Review 1981* (Ottawa, 1981), pp. 45-59

[12] Science Council of Canada, *The Proposal for an Intense Neutron Generator: Report No. 2* (Ottawa, 1967)

[13] J. Ronayne, *The Allocation of Resources to Research and Development — Report to the Australian Science and Technology Council* (Sydney, 1979), p. 55

[14] Royal Commission on Government Organization (6 vols., Ottawa, 1963), IV, p. 220

[15] *The Urgent Investment: A Long-Range Plan for the National Research Council* (Ottawa, 1980), p. 5

[16] *A Five-Year Plan for the Programs of the Natural Sciences and Engineering Research Council* (Ottawa, 1979), p. 9

[17] G.C. Allen, 'Industrial Policy and Innovation in Japan', in C. Carter, editor, *Industrial Policy and Innovation* (London, 1981), p. 72

[18] T. Dixon Long, 'The Dynamics of Japanese Science Policy', in T. Dixon Long and C. Wright, editors, *Science Policies of Industrial Nations* (New York, 1975), pp. 133-68

[19] N. Sullivan, *The Japanese System of Technology Promotion — An Overview* (Canberra, 1977), pp. 2-4

[20] OECD, Reviews of National Science Policy — Japan (Paris, 1967)

Chapter 8

[1] Science Policy Programming Services, *Science Policy in Belgium* (Brussels, 1979), pp. 24-5

[2] ibid., p. 17

Chapter 9

[1] OECD, *Science and the Policies of Governments* (Paris, 1963), p. 5

[2] OECD, *Science, Growth and Society: A New Perspective* (Paris, 1971)

[3] OECD, *Technical Change and Economic Policy* (Paris, 1980)

[4] W.R.A. Wyllie, *Australia and the Western Pacific Region ... The Future*, Paper delivered to the Victorian Export Action Committee, Department of Trade and Resources, Melbourne, 10 June 1981

[5] Central Policy Review Staff, *Social and Employment Implications of Microelectronics* (London, 1978), p. 3

[6] J. Sleigh, B. Boatwright, P. Irwin and R. Stanyon, *The Manpower Implications of Microelectronic Technology* (London, 1980)

[7] J. Rada, *The Impact of Micro-electronics* (Geneva, 1980)

[8] G.W. Rathenau, *Report of the Advisory Group on the Social Impact of Micro-electronics* (The Hague, 1980)

[9] P. Stubbs, 'Technological Change in Australia: A Review of the Myers Report', *Economic Record*, Vol. 57 (1981), pp. 224-31

[10] *Report of the Committee of Inquiry into Technological Change in Australia* (4 vols., Canberra, 1980), Vol. 1, p. 73

[11] OECD, *The Current State of Science and Technology Policy* (Paris, 1980), p. 6

[12] Ministry of Education and Science, *Sector Councils for Science Policy* (The Hague, 1978)

[13] J.G. Crowther, *Statesmen of Science* (London, 1965), p. 251

Index